Agricultural Development: Soil, Food, People, Work

Agricultural Development: Soil, Food, People, Work

by CHARLES E. KELLOGG

Editorial Committee

M. D. THORNE, University of Illinois, chairman

W. V. BARTHOLOMEW, North Carolina State University

R. D. BRONSON, Purdue University

Managing Editor: RICHARD C. DINAUER

Assistant Editor: DEBORAH F. COONEY

1975

Published by

Soil Science Society of America, Inc.

Madison, Wisconsin, USA

Second Printing 1977

Soil Science Society of America, Inc.
677 South Segoe Road, Madison, Wisconsin 53711 USA

Library of Congress Catalog Card Number: 75-5204

ISBN Number: 0-89118-763-4

Printed in the United States of America

186 3317

Contents

Foreword

One of the most important prerequisites to the success of any venture is to know the resources that are involved. Dr. Charles E. Kellogg is admirably qualified to write on agricultural development. He has studied the genesis, morphology, classification, and use of soils across the world landscape and analyzed all things with which his several senses placed him in contact. In fact, Dr. Kellogg has spent a lifetime studying—first and foremost—the soils of the world, and then the agricultural, industrial, and sociological systems within which man must wend his way. Gathered in this book are the recollections, facts, and philosophies of a man who has seen much of the world; he has observed it personally and through the visions of others.

This book will be of value to a wide spectrum of readers. The agriculturist will appreciate it because it can be a real stimulus to those searching to understand the agricultural enterprise on a world-wide basis. The lay reader will enjoy the way Dr. Kellogg has been able to bring together, perhaps for the first time, facts and philosophies of the farmer (the cultivator) as he functions in agricultural systems from the most primitive to the most complex. The author has made an attempt, however difficult, to separate the text into chapters. He has succeeded but only to the degree that a world-wide subject steeped in essentially all the natural and social sciences will permit; because of this he has admonished the interested reader to scan the entire book first before searching out the detail.

The successful development of several major cities and countries continues to be closely related to the inherent productivity of the soil. The region surrounding the Nile River and our own country are prime examples of highly successful civilizations that are based on excellent soil resources and the agriculture this fosters. Dr. Kellogg has captured the essence of this as he discusses many of the factors that have made the developed countries great and promise to make some of the developing countries reach equal pinnacles of success. He is optimistic, however cautious, about agricultural development in the future because, as he says, "the basic principles of agricultural development do not change significantly with time or place."

The Soil Science Society of America is priviledged to sponsor and publish this book—a product of one of its outstanding members, and a national and international leader in pedology, a basic study in agriculture.

31 March 1975   CHARLES F. ENO, *president*
                                        Soil Science Society of America

Preface

This smallish book is intended for busy people who need a view of the basic principles of agricultural development and the problems and opportunities for it in a complex world.

Because nearly every aspect of the subject is related to nearly every other aspect, I hope the interested reader first skims the whole book.

Since many natural and social sciences and technologies make substantial contributions to this subject, I know well that others are better informed in several of the disciplines than I am. Yet if no one departed from his own special field of inquiry, we should have no broad treatments of any complex, world-wide subject.

Few could have had more great teachers than I, not counting the many I have met only in books. In Michigan, L. R. Schoenmann first showed me what soils mean to forestry and farming, and V. R. Burton encouraged me to apply soil science to highway design. I learned much from discussions at the University of Wisconsin with scholars like Emil Truog and George S. Wehrwein; at North Dakota State from H. L. Walster, L. R. Waldron, and others; and from the splendid seminars in the U. S. Department of Agriculture in the days of Henry A. Wallace and M. L. Wilson. C. F. Marbut, my predecessor as head of the USDA Soil Survey, expanded my view of the world and my first chief in USDA, Henry G. Knight, was a good teacher in research administration. There was Bushrod Allin and many others in the splendid staff of the Bureau of Agricultural Economics. I learned much from John D. Black of Harvard; R. M. Salter, USDA; J. C. McAmis and H. A. Morgan of TVA; Tom Wallace of the University of Bristol, G. W. Robinson of Bangor, and others in Britain; C. H. Edelman at

Wageningen in The Netherlands; and C. F. Charter of the old Gold Coast.

Many are still working on my education, including T. W. Schultz of Chicago and Charles M. Hardin of California; Richard McCardle and his successor, Edward P. Cliff, Ralph W. Phillips, Max I. Witcher, and many others in the USDA; Georges Aubert in France; E. Walter Russell of Britain; S. P. Raychaudhuri, who showed me much of India; René Tavernier of the University of Ghent and F. Jurion, also of Belgium, who showed me the Republic of Zaire (Belgian Congo); V. Ignatieff of Canada; Tom Walsh of Ireland; Norman Taylor of New Zealand; L. J. H. Teakle of Brisbane, Australia; and others.

I also owe much to the hard work and to the both cool and hot debates of the great staff of the National Cooperative Soil Survey and their colleagues in other agencies, especially in the land-grant universities, over the past many years.

The U. S. Department of Agriculture gave me many opportunities for travel abroad for which I am grateful. So did several other governments and both the Rockefeller and Ford Foundations.

Besides my professional associates and guides I have learned from the people on the land in both rich and poor countries. Some had very large individual holdings, others had tiny ones. Some worked in collectives and cooperatives. Others were shifting cultivators and herdsmen. Some lived on their holdings, others lived in villages, and still others were nomadic and moved with the seasons. Many combined farming with forestry, fishing, or other occupations associated with farming.

Without the great help and encouragement of my wife, Lucille J. Kellogg, this book would not have been possible.

With all the ways of growing food and industrial crops we have some difficult problems with words. The equivalent of the word *farmer* in other languages can mean *estate or plantation manager*. Actually we shall be concerned with landless laborers, tenants, and sharecroppers, as well as owners and managers. The only term I know that includes all of those who work the soil to produce crops is *cultivator*.

Also we find wide variations in the state of development of countries and of parts of countries. It is common practice to speak of the "developed countries" and the "less developed countries," or "ldc's." For those ldc's that are moving forward the terms "developing countries" or "newly developing countries" are used. These are all vague terms with no clear boundaries. If 12 well-travelled scholars of agricultural development were asked to put all countries into one group or other we should likely have 12 different lists. India, for example, has had a great culture for a long time. Large parts of other countries have no indigenous written language. Because a new combination of practices spreads fairly rapidly in India, for example, is not to suggest it would go forward in the same way in other countries now, say the Republic of Zaire (Belgian Congo).

What people may do well in one part of a less developed country may have little relevance for people in other parts with different cultural backgrounds. [See for example, Jean Pierre Hallet's *Congo Kitabu* (1965) or Colin Turnbull's *Forest People* (1961) on the pigmy people of Africa.]

Probably the worst collective term for the ldc's is the *third* world, perhaps in contrast to the *free* and the *communist* worlds. Such a classification reminds me of the example of a vague one used for illustration by my old friend C. F. Marbut—a classification of houses into "little houses," "brick houses," and "red houses." When we use the term *developed countries* we must be aware that they probably have undeveloped areas, even large ones in some.

Although this book deals mainly with less developed countries, some examples of practices are also taken from developed areas. Comparable practices from developed areas can be used, in proper combinations, in newly developed areas with similar soils and climates—as local education and infrastructure permit.

Another very difficult semantic problem is the common confusion of the words *farming* and *agriculture*. As agriculture develops, many of the jobs formerly done on farms are taken over by nonfarmers living in villages, towns, and cities. Such agricultural workers make the farm machines, irrigation pipe, chemical fertilizers, and all the rest. They now do most of the processing and transport work. When a farmer buys a machine he is really paying for nonfarm labor. And this labor is a vital part of the whole agricultural system. In the United States, for example, only about 10 to 15% of those working full-time in agriculture actually work on farms. In other countries the percentage on farms may go to well above 50% of the workers in the whole country, whereas in the United States about 30% of the national labor force works in agriculture but only 4 or 5% of the total labor force works on farms.

We must also beware of some of the general economic terms, used in comparing countries, such as *gross national product,* (GNP), without considering how it is calculated in different countries and how well it is distributed among the people. In fact GNP can be concentrated among so few as to lead to general poverty, injustice, and even revolution. In many countries, including some "developed" ones, only a small minority have access to soils responsive to management for farming. Reasonable programs of land reform are essential for justice and stability. After all, who does *own* a country?

Critical questions of economics arise all along the way. And it makes a difference who's economics is used. For such quandries I urge the little book by Joan Robinson on *Economic Heresies: Some Old Fashioned Questions in Economic Theory* (1971).

Despite the overdrawn warnings, there is no general crisis in food production or deterioration of the rural environment, except as wars, persecutions, and denials of opportunities are allowed to continue.

Nor is there any simple panacea for success in agricultural development. Fertilizers, water control, new varieties, pesticides, multiple cropping, and so on all have their place in combination with one another, as adapted to the local kinds of soil. But none is a panacea.

Nor must we expect sudden miracles. Effective agricultural development requires much *hard work* by many people having a wide variety of skills.

We can be cautiously optimistic. The basic principles of agricultural development do not change significantly with time or place. But the problems differ from place to place and at different times. Diagnosis is the first critical need. Success will not come in only a decade or so. The problems of food and agricultural development require the work of many skilled people, not simply the cultivators alone.

March 1974
Hyattsville, Maryland

CHARLES E. KELLOGG

1

Introduction

This chapter is intended to lay out the broad principles for successful farming that apply generally, with local adaptations, of course,
and to illustrate these with examples from both developed and less
developed countries. Then too, some glimpses of the history of
farming and how it reached its present state are helpful. The main
thrust is to introduce essential concepts to those readers working
in the less developed countries.

Many people are deeply concerned about the future food supply of the world. Increased supply means more agricultural development—both larger total areas in farming and higher output per
worker and per hectare. Then too, for farms to produce more
crops, those who work in the fields will need more tools, machines,
fertilizers, and other services within the broad field of agriculture.

There is no panacea. Only in proper combinations can our
soils and practices produce the many kinds of food crops, animals,
and industrial crops that will be needed. Since the world has a
very wide variety of kinds of soil, climates, water supplies, local
skills, and markets, a host of proper combinations of practices
must be understood and used.

Since 1930 great progress has been made in breeding better
varieties of crops, for making better and cheaper fertilizers, in
developing systems of water control, harvesting, and land clearing,
and also for the storage and processing of farm products.

More recently we have heard much about the "green revolution," presumably due mainly to the new higher-yielding varieties
of cereal grains—wheat (*Triticum aestivum* L.), rice (*Oryza sativa*
L.), and maize (*Zea mays* L.). These and other products of our
plant geneticists are very promising—but *not by themselves*. The
new, potentially high-yielding cereals were bred and selected to re-

spond to good water management, to weed and other pest control, and to the appropriate kinds and amounts of fertilizer to supplement the plant nutrients in the soil. Without these additional practices, especially water control and fertilizers, the old varieties do as well or even better than the new ones on most old fields.

As we shall develop more fully in this and the following chapters, each aspect or practice for growing crops depends on all the others—the kind of soil, fertilizers, tillage, water control, pest control, kinds and varieties of crops, forage, and trees,—along with the technical, educational, social, and economic services for the cultivators.

PRINCIPLE OF INTERACTIONS

The key axiom in agricultural development is the "principle of interactions." On a field, for example, the cultivator must have a good combination of many factors suited to his soil, to have even an "average" yield. Each part of a combination of practices interacts with, and influences the effectiveness of all the other parts. Rarely do notably higher yields or higher quality crops result from a change in one practice alone. Without fertilizer, for example, a new variety or irrigation alone can be disappointing. Then too, even if the cultivator uses the proper combinations of practices for his soils, he may be only a little better off without adequate storage and markets.

Those working to improve production and human welfare on existing farms, or to bring new soils into use for crops must continually recall that each of the many factors relates to all the others within the combinations that lead to success. Obviously this principle of interactions applies not only on the farms but also within the community or trade area that makes services available for farming.

Also in newly developing countries, the development of agriculture to high levels is bound to be a fairly slow process since neither all the essential skills nor the "infrastructure" can be introduced at once. Any functioning community, whether agricultural village, city, or country, is dependent on the continuous operation of basic services—transportation, education, health care, banking, communications, etc. As in the example of improved crops and yields, each section of the infrastructure influences the workings of the others.

Since each of the developed countries got its start toward economic development when farming became efficient enough to produce a surplus over local needs, it may be helpful to look briefly at the early history of agriculture, the principle of interactions, and the other major themes to be covered in more detail subsequently. In several countries savings for investment were accumulated from fishing, forestry, and mining. Yet farming, according to Adam Smith and others, was the principle source of savings for most countries prior to the "age of science" and the great European discoveries.

Despite man's long struggle for food, our Earth actually has abundant soil resources. Man must find the responsive soils, those that give good yields with a proper combination of practices, and learn how to use them on a sustained basis. This has been difficult to do because different kinds of soils, even those with similar high potential, respond to quite different combinations of practices and plants.

The early settlers from England into what is now the eastern part of Massachusetts in the United States had very low yields from the seeds they brought with them. Although the soils were similar to many in England before cultivation, they were less fertile than those that had been farmed well in England since Roman times. The local Indians showed the settlers how to grow maize with a fish buried near the seed, presumably to the side and below the level of the seed, so that the roots would start in the soil alone and grow into the fish without salt damage.

Then each family was given a parcel of land by lot and required to grow as much maize as it could. Later, however, each family wanted a *specific* plot (about one-half hectare) so they could have the advantage of their own improvements of the soil. Then cattle were brought, and each family needed a larger holding to insure their own food and some to sell (Bradford, 1856).

Shifting cultivation

We shall take a look at shifting cultivation because many people still practice the system. The way it works illustrates both how

4

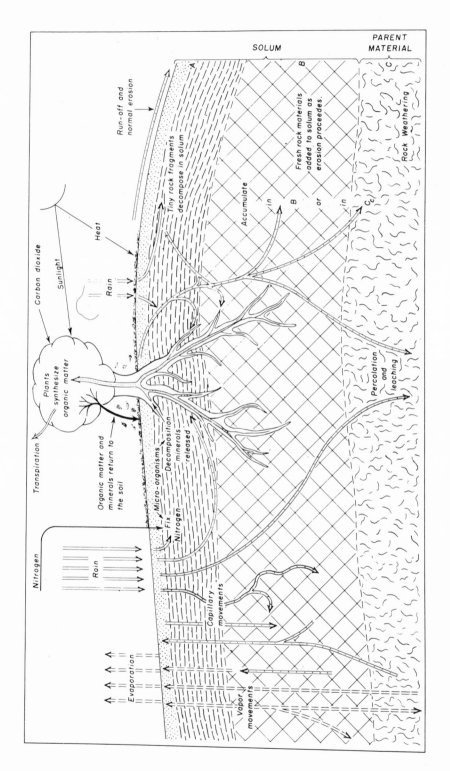

many soils were formed and how soils respond to modern management.

In the forested areas early man learned that trees, cut and burned, rejuvenated soil productivity. A large number of systems were developed and some are used today in areas remote from transport.

The basic principle is simple (Figure 1). As trees grow, they take nutrients from the soil, from dust on the leaves, and from nitrogen fixed by electrical storms which falls on the soil and plants during rain. Also some ammonia nitrogen returns in rain falling through smoke. As leaves, fruits, stems, and so on fall from the trees they can be partially eaten by large and small animals. The remainder and the animal waste is decomposed by the microorganisms and released back to the rooting zone of the soil. Once a nutrient ion gets into this nutrient cycle there is little escape from it. Obviously few of the nutrients in the living matter can be leached beyond the root system. In the humid tropics the circulation is very rapid even if the total supply is not great. Then when the trees are cut and burned the stored nutrients are released to the soil.

In most cases not all the wood is burned completely, but decay is nonetheless rapid. Where the growth is poor, sticks and small logs may be gathered from adjacent soils not suited to farming to add to the total supply of nutrients. The fertility of some fields depends greatly on wood brought in from nonfarm areas and some on compost made from leaves and branches. In Assam the plots are laid out in beds and the wood is covered lightly by thin sods before firing (Figures 2 and 3). An "expert" from the outside showed the local cultivators how to grow potatoes (*Solanum tuberosum* L.) in the usual way as done elsewhere. But he harvested almost no potatoes because of wireworms (*Haemonchus contortus*)!

Actually any outside expert must be cautious about changing a local procedure until he understands why it works as well as it does. Rarely do the people know why their methods work, although they may give unbelievable reasons. But once the real reasons are learned the scientist may then be able to improve the system greatly (Jurion and Henry, 1969).

FIGURE 1

This simple chart shows how plants take in nutrients from air, soil, and water which they then synthesize into organic matter under the power of sunlight in the green parts. Water moves up and down in the soil, through it, and off the surface.

FIGURE 2

Here in Assam, India, the soil is resting after cropping. Cultivators place wood and twigs on the soil surface and cover lightly with sods, in preparation for burning.

FIGURE 3

In Assam, India, the beds are burned. The soil is thus enriched with minerals; weed seeds and pathogens are killed, and the soil is made friable.

Shifting cultivation is the general term for systems of farming based on this process. Yet nearly every language has or has had a term for it, such as *milpa* in South and Central America, *kaingin* in the Philippines, *rai* in Vietnam and Thailand, and *jhum* in Assam, India. A search of all oral languages would disclose many, many more.

Bush fallow is a specific equivalent for the basic process, if the reader realizes that the word *bush* is commonly used for all vegetation hard to walk through. A broader term is *natural fallow* to include systems based on excellent elephant grass (*Pennisetum purpureum*) as the fallow plant instead of trees. *Swidden* is an old English term for *burned field,* but not equivalent to shifting cultivation.

Some use the term *slash and burn,* especially where the trees and shrubs are cut and burned, and the soil farmed until plant nutrients are exhausted. Then with severe erosion on sloping soils the trees do not return without systems to control the runoff and erosion.

Most use the term *shifting cultivation* specifically for a system of cropping for about 2 to 6 years after cutting the fallow forest. The soil is subsequently returned to forest for about 8 to 15 years, depending on the soil, for restoration of productivity. This restoration is not of soil fertility alone, but also of the structure of the soil. Actually the growing roots give the soil a kind of tillage. While a little root grows to a big one, much soil is moved. Then too, during the fallow periods weeds and their seeds are destroyed along with many insect pests. Dry seasons in some regions and cold winters in others have therapeutic effects on weeds and other pests. Such effects are lacking in the humid tropics. In Figures 4, 5, 6, and 7 are examples of stages of shifting cultivation in a regular corridor system.

The vast literature on shifting cultivation is widely scattered. Some authors include the social consequences of the system only, without realizing that the potentials of specific systems vary widely among different kinds of soil. Bartlett (1955–1961) developed a large bibliography on the use and misuse of fire in tropical regions. Nye and Greenland (1960) published a book with examples mainly from West Africa. De Schlippe (1956) has one dealing with the Zande peoples in the northern part of central Africa. Jurion and Henry (1969) deal with the improvement of the Bantu system

FIGURE 4

After 9 years of lying fallow under umbrella trees (*Magnolia tripetala*) of the tropical rain forest, the soil has been stirred by the roots and cropping can soon begin. (Low growth has been cut for the photograph.) North of Yangambi, Zaire.

FIGURE 5

The forest is cut and will be burned for cropping to follow. Yangambi, Zaire.

FIGURE 6

This is the third stage of cropping, following clearing, the planting of bananas (*Musa paradisiaca*) and upland rice, and after maize harvest. Yangambi, Zaire.

FIGURE 7

This is the fourth stage of cropping, after a harvest of bananas and cassava (*Manihot esculenta*). The trees will return under the shade of second-growth cassava. Yangambi, Zaire.

of shifting cultivation and give research results, along with those from other kinds of farming.

This kind of farming was common in the early settlement of northern Europe after the retreat of the glaciers some 10,000 years ago. In the old epic poem of Finland, *The Kalevala* (Crawford, 1888), the singer tells us:

Osma's barley will not flourish
Not the barley of Wainola
If the soil be not made ready
If the forest be not leveled
And the branches burned to ashes.
. . . .
Never will the earth unaided
Yield the ripe nutritious barley.

Shifting cultivation was practiced in the northern part of Russia, where fuel and water were abundant, before the famous black soils in the treeless plains—the Chernozems—were regularly used for crops. In the early years, before steel pipe, water was scarce in the Chernozem belt and there was danger of invading cavalry.

Gradual improvement

Today the world has abundant resources for producing food, fiber, timber, and a wide variety of industrial crops. Because of the accidents of early climatic change, migration, and social change, and in light of enormous changes in agriculture and its technology, well over 50% of the potentially arable soils of the world are not being used to any large extent for farming. And a large part of those being used could have far larger harvests.

Early man was primarily a food gatherer. He took whatever wild plants, fish, and land animals the unassisted natural system made available. Essentially all of the population was engaged in this effort. Certainly life was risky at best.

The early development of agriculture was not a steady process from one source. No doubt it had many independent early beginnings in several parts of the world. Some of these appear to be related because of similarities. But for any start to survive, certain basic principles must have been learned by trial-and-error or by accident and then applied. So there are likely to be some notable similarities among systems developed independently on similar kinds of soil.

Gradually some people learned how the soils could produce more. Seeds of some of the better food plants were collected and planted. Greater harvests justified the effort. Some animals were domesticated. A clever person, perhaps by accident, learned to reproduce olive trees (*Olea* sp.) by planting some cuttings. And so we might go through a host of items that made farming more efficient.

Lord Zuckerman (1971) has pointed out that natural science really had its beginning in farming and in the earliest villages, which included potters, weavers, and both stone and metal workers. Actually the social sciences also began in these very early villages. As Darlington (1971) pointed out in his great book on *The Evolution of Man and Society,* the development of skills in farming related closely to man's development of social life, trade, and government.

Long before historical times, perhaps about 9,000 B.C. in the Near East, and in Northern Europe around 4,000 B.C., people made tools for cutting brush and trees, for spading the soil, and for other essential jobs in raising food crops.

Many improvements in farm implements were made during Greek and Roman times. Columella—a Spanish Roman—wrote his great book *Husbandry* about 60 A.D. and it remained the principal authority on farming in Western Europe for several centuries. Ibn-al-Awam, a Moorish scholar, wrote his *Book of Agriculture* in the latter part of the twelfth century. From it we know that the Arabs had made progress beyond practices of Columella's time.[1]

New emphasis was placed on both basic science and exploration during the late thirteenth century, about the time that Marco Polo made his famous journey to China. By learning how plants grow in the soil, cultivators increased both soil productivity and harvests. Scientific knowledge grew quickly in the next centuries, laying the basis for the industrial revolution of the eighteenth century. Ironically, Malthus wrote his pessimistic essay on population, just at the close of the eighteenth century before the great application of science and technology to farming. The electric dynamo and the steam engine came in the nineteenth century.

---

1. Many translations of Columell's book are available in English, but only Spanish and French translations of Ibn-al-Awam's book exist now.

THE NATURE OF AGRICULTURE

Up to the industrial revolution, which was based on scientific and technological innovations, the words *farming* and *agriculture,* and their equivalents in other languages, had been almost synonymous. In the seventeenth and eighteenth centuries most farm tools were made on the farm or in the village blacksmith shop. A high proportion of the farm products were processed for marketing on the farm or in the village.

Trade and commerce were already concentrated in the larger cities. Since all heavy transport was by ship, nearly all of the cities were near the margins of oceans or navigable lakes or rivers. Incidentally these were generally poor places for extensive building. During roughly the past 3 million years, the water levels in oceans and rivers have changed drastically—low when the ice was piled on the land and high when it melted. As ocean levels rose, the sea cut new shore lines and deposited sediment. When the water receded, the streams washed out the sediment unevenly. Similarly the streams and lakes fluctuated. The result of three such changes in the past 3 million years was a complex of clay, silt, sand, gravel, and even muck where most of the port cities were located. Thus good foundations for roads and buildings are now costly.

Steam put great power in the workman's arm, but neither could the steam be made nor the workmen live far from the factory. Thus began the great building in the port and trading cities. To make matters worse, as the jobs formerly done on farms were moved to the cities—jobs to make machines, chemicals, and all the other farm inputs, jobs for processing farm products—agricultural workers had to move from the farms to the cities to find this work. Thus, the jobs in the industrial sectors of agriculture grew and those in the farming sector declined.

These changes accelerated greatly in the twentieth century. Before World War I, for example, most Americans lived in the country, whereas after that war most lived in the city. After 1935 the process quickened even more. For example, in the United States during the late 1930's and the early 1940's nearly all the enormous handwork in cotton growing was eliminated—the thinning, weeding, and picking. A cotton farm with 75 families formerly now has a few tractor drivers. No one touches the cotton. But the farmer pays a great deal for "city" labor in the price of machines and chemicals.

The handwork for growing many other field and horticultural crops has been greatly reduced and the trend continues. The outlook in the United States and other technologically advanced countries for any farm production requiring hand labor, except for luxury items and services, is dim indeed; today only about 15% of the labor force in agriculture, or about 4 to 5% of the national labor force, lives or works on farms.

Since these changes in the nature of agriculture came gradually many people have not sensed them during their adult lives. People say, for example, that during the past 100 years "agricultural employment has decreased drastically in the advanced societies." This statement is false. The decrease has been in the farming sector. In the industrial, finance, and service sectors of agriculture, employment has increased. Probably total agricultural employment has not decreased so much as people have been led to believe.

Looking at farming alone, the purchase of machines and the like requires capital. The change can be called a switch from "labor intensive" to "capital intensive." Yet in making the change the farmer or cultivator is paying largely for labor in a different way.

Thus, the use of the term *agriculture* for the farming sector alone is misleading. Many of the greatest agricultural advancements have been in the industrial sector. For example, the reduction of food waste between the unharvested crops and the kitchens of consumers, thanks to the availability of electric power on farms and modern processing facilities and transport, has been truly enormous in the developed countries.

To add to the confusion, the word *rural* connotes only farming to many people. Yet, to be fully successful, every farmer needs a village or town nearby with storage and marketing facilities, supplies of machines and chemicals, and several other farm and social services. People no longer need to depend wholly on railroads for transport.

The confusion between "farming" and modern "agriculture" has misled some general economists. For example, Sir Eric Roll in a splendid book, *The World after Keynes* (1968) writes about the "political weight. . .of the rural vote," which he assumes to be less than 4% of the national labor force. But this is only the farm part. I do not know the percentage of the British labor force in agriculture but it is certainly very much higher, probably around 20 to 30%, considering food processing and farm input manufacturing, such as farm chemicals and machines, which are also important export items for Britain.

This sort of error leads to others, such as contrasting "agricultural development," in the sense of farming only, with "industrial development," which may or may not include the large industrial sector of agriculture. This is clearly evident in many recent discussions of the progress of development in the less developed countries (ldc's). (See Ronald Robinson, 1971, page 68 *et seq.*) In this discussion the essential industrial sectors of agriculture are not mentioned. In developing countries these should obviously have a high priority for agricultural development, along with increased nonagricultural industries and employment.

Modern agriculture is a very large economic enterprise requiring many factories as well as farms and large numbers of people with wide ranges of skills. To omit vital parts of the system and concentrate on farming alone is to preclude significant agricultural development (United Nations, 1963).

Herein lies the basic reason for many failures of "rural development" programs in the less developed countries as guided by outside technical assistance focusing exclusively on farm improvements. In many of the existing villages, part of the people live by making hand tools, carts, cloth, shoes, and so on. What about them? Also if the new inputs and processing are developed in distant cities what about jobs for the cultivators and other villagers who will become superfluous? In some schemes no one even thought about any storage or marketing facilities for the increased product. It is not too difficult for a good farm expert to increase yields several fold with fertilizer, good water control, superior varieties, and the like. Such successes without storage and marketing have lead to disastrous drops in local prices.

Town-and-country

It would be unfortunate if people in the less developed countries tried to follow the same path toward agricultural development that historically occurred in the developed countries, with their overcrowded port cities and the continuous migrations into them. In new developments people can make viable towns out of some of their existing villages. These can be served by highways and do not need to be on navigable streams. It would be a great pity if the agricultural industries were all or even primarily located in the older big port cities. If this should happen and the farming improves, we would have every reason to expect long lines of migrating hopeless, unemployed farm and village people going to the big cities, perhaps to die in the streets. The people need jobs as well as food.

Recently Jonathan Power (1972), in *The Last Thing Africa Needs is Cities,* has made this point very clear as it related to the cultivators accustomed to village life in the country. If the blacksmith, for example, moves to a large port city, his life and that of his family are worse; so are the lives of the villagers. Because he has left, they have no blacksmith to make or to repair the essential tools to which they are accustomed.

The country villages or towns can and should have employment for those former villagers and cultivators who will not be needed on farms in improved farming systems. With the essential common services for farming and for at least part of the industrial sectors of agriculture, the local town would also have the infrastructure for industries based on other local resources. The common contrast between "rural population" and "urban population" suggests artificial divisions that are no longer necessary and certainly not desirable for either the crowded city dwellers or for culturally deprived and unemployed rural dwellers.

"Rural development" can be an excellent program if the word *rural* is meant to include all natural resources within the area, not simply the soil and water for farm production, and all the essential town services. Actually, the term *town-and-country* development is much more expressive of what is needed and has in it a place for farmers, businessmen, and a wide assortment of skilled labor (Kellogg, 1971). (*See* Chapters IV and IX.)

Because this change in the nature of agriculture since about 1860 was gradual—the declining numbers of workers in farming, the increasing numbers of workers in the industrial and service sectors of agriculture, the need for viable country towns—it has been missed by some. A few planners are even advocating almost revolutionary schemes of land reform under which cultivators would use "labor intensive," inefficient systems, which could lead to poverty.

In many countries, the cost of planning towns properly has become enormously high because of failures to provide ceilings on land prices. The great increases in some land prices have made a few people wealthy for no contributions whatever to society.

Food and jobs

Even with potentially adequate food and industrial crops produced, people need to have money to buy them. In many village areas today, it is assumed that every family has an inherent right to land—enough to grow their food. As agriculture develops, however, a smaller proportion of the people in agriculture will actually work on the land. So this valuable old principle of rights to land must somehow be changed to rights to a job. In any society those who do no useful work become social charges on the others. And hopefully the job for an individual will not be far away from the people of his own local language and culture.

One must begin with the facts

Before large investments are made of funds and labor for farm improvement or for new farm development, the soils must be looked at carefully. Their potential as nature made them may have little relevance. The question is rather how well they respond to various achievable levels of technology and management. Actually the soils used by most of the good commercial farmers of the world are much more productive now than they were as nature built them. Many kinds of soil originally low in plant nutrients have been given applications of lime, manure, and fertilizers over the years and water has been controlled to avoid more than normal soil erosion until these soils are much more productive than when first used for crops.

Appraisals are needed for dependable water supplies, potentials for forestry, fishing, mining, hydroelectric power, and other economic enterprises as well as soils. In fact, few successful and rewarding farm areas can pay out of farm income alone all the costs for the essential infrastructure—transport, education, health, and other common services. In nearly all thriving farming areas that I know about these costs are either widely shared by several productive enterprises or they are at least partly subsidized by government. Usually this second course can be justified only for a start.

Later we will show that the aim should be a potentially successful trade area with marketing facilities for all, including the processing and marketing of farm products and supplies of essential farm inputs, with many produced locally. As the farming improves the superfluous cultivators can have jobs within the area. For the world this means many thousands of new inland towns.

During the past 25 years much has been learned about the location of such towns to avoid the horrendous problems of unstable roads and buildings, pollution, flooding, and so on in the old port cities and their suburbs. Actually these towns can have good schools and colleges, good stores, and cultural activities for all within the trade area. Industries can be located to avoid much of the potential air and water pollution. People can avoid a choice between employment and clean air and water in their cities. There is no need to cover the soil so completely with pavements, parking lots, and buildings that the falling rain cannot soak into the soil so that little streams become raging torrents after every rain or thaw.

The principle of interactions applied to soil and farm improvement

As already pointed out, few kinds of soil are productive naturally for the highly bred and selected crops grown for food and industrial uses. Yet many are highly responsive to special combinations of practices, fitted to one another and to the *local* kinds of soil. Each practice interacts with the others and with the many characteristics of the local kind of soil, which, in combination, make it of a specific type.

Every hectare of soil in the world that gives a good harvest for the inputs of labor and materials used has a combination of practices that provides for the following: (i) a balanced supply of nutrients in the soil for the kinds of plants to be grown; (ii) adequate amounts of both water and oxygen in the rooting zone as the plants require them; (iii) kinds and varieties of adapted crops or trees that have the genetic potential to respond to the most nearly ideal soil conditions it is practicable to develop; and (iv) suitable measures for controlling pests, including weeds. The omission of any one of these variables can nullify the potential effects of the others (Kellogg, 1962).

Besides the well-nigh universal sets of practices to achieve those four objectives, additional practices may be needed to make otherwise responsive soils suitable for farming: protection against mountain torrents, the sea, strong winds, fire, and other environmental disturbances.

Besides these interactions in planning the use of a field, the cultivator must think how to use his several fields, commonly with contrasting kinds of soil, so they fit into a complete system for the use of his holding. If he has livestock, he will select those soils on which it is easiest to control water for his field, horticultural, and industrial crops, and use the more sloping and stoney soils for pasture or harvested forage.

Then too, what he can do depends on his skill, education, credit, markets, and both the price and availability of such inputs as fertilizer and tools. Thus the interactions between the alternative potentials of his farm and the services in the local village or town are highly important. Without a nearby town he is limited to subsistence farming, as are millions of shifting cultivators and other cultivators, plus perhaps, an industrial crop, such as cotton, that can be carried long distances on the head or on the back of an animal. As farming does improve, the need for a fair-sized village or town becomes increasingly important. Only a part of the whole agricultural system takes place on farms.

This principle of interactions is vital, and a main theme of this book. One of the great problems in agricultural development is the selection of the right kinds of practices and resources for the desired result. A combination of practices that may work well for one type of farming or forestry may work badly in a different combination or in another area. No practice should be promoted or condemned because of its success or failure in one area. Deep plowing, for example, can be good or bad in different situations. Composting can be an excellent practice or an uneconomic and wasteful one. Many forms of terracing work well in proper combinations of practices but actually stimulate soil erosion in others. Clear-cutting is an excellent forestry practice or a poor one depending on the stability of the soil and the kinds of trees that come in as second growth.

These are only a very few out of a huge number of possible examples. No practice used by successful farmers or foresters is universally good or bad. One needs to size up the whole situation and understand as best he can the host of interactions among practices, the soils, the available services, and the local culture. Many unscientific, and even anti-scientific, "universal" statements are now being promulgated about the influence of certain practices on the environment, such as improper fertilizer use. Scarcely a one can be followed exclusively for application to any significant percentage of the real farms or forests of the world, or to the potential ones.

Where to start

Only a few very general principles can be suggested here about where to start in an area or country with an obvious lack of rewarding opportunities for work. Usually this means ways of helping cultivators achieve an increase in their living standards, which requires increases in productivity of both the hectares and the people.

Certainly one needs to appraise, generally at least, the kinds of soil and their potentials for producing both food and industrial crops, the potential water supplies, and resources for forestry, mining, hydroelectric power, and grazing. A clear notion is needed of potentials for transport and other essential services.

Most important, and commonly neglected, is to get to know the people, to find out what they want most. What Europeans, Americans, or others might want most if they were to live in the area may be a poor guide indeed to what the local people themselves want. Unfortunately, many in technical assistance have not learned the skill—apart from language differences—of communicating with the peasant people. Simply because they do not read does not mean that they are ignorant nor without cultural values. If the advisor intends to help them he must avoid giving any impression that he thinks himself to be above them. To do so shows both lack of understanding and bad manners. The advisor will earn no trust from the people this way.

Then too, one needs to be aware that words can suggest quite unlike concepts to different people. And, of course, translations, which most people must use, can make matters worse. Our word *freedom*, for example, may come out in another language as *license*—failure to consult the others in the family or in the village— which is anathema in some other cultures.

It is important to understand the social system in which the cultivator and his family live. What gives status? How are both the individual and collective decisions made? This also requires some clear idea of how the cultivator relates to his government— national, provincial and local—both directly and indirectly, and to local law. It may be through the so-called natural leaders or through clans, guilds, or other associations (Andreski, 1968).

Besides current living patterns, the skills of the cultivators and other people must be evaluated, otherwise there is the danger of suggesting systems too sophisticated for them. We have already emphasized the principle of interactions. In some areas one can introduce immediately sets or "packages" of practices. This was done successfully in many Indian villages during the 1960's.

Yet if one were to try to introduce the same practices with more primitive cultivators, they might be very confusing. Rather, one might start simply by introducing a cash crop. If they were near a mine or town, this could be a food crop. If not, an industrial crop might be a better choice. With an initial success, one might introduce compost, runoff control, or fertilizers.

Some sort of secure land tenure is nearly essential for a cultivator to be expected to make savings out of his meager income to improve the potential of the soil he cultivates. Societies have used several ways to achieve this security. Many schemes were inherited from a dim past. Most of these schemes now depend on the government and may give rise to great opposition from resident or non-resident landlords or an elite group who may fear for their traditional status and incomes should the cultivator become a free man, able to get the benefits of improvements that he makes in the soils he cultivates (Andreski, 1966). In some situations a powerful few oppose developments that would give great increases in labor income to the many (Frank, 1969).

Not only reasonable plans for land tenure, but other programs for improvement must consider the laws of the area—constitutional law, statutory law, and the common (or tribal) law. Despite certain eternal verities, these laws vary among countries and among parts of a single country. In the United States, for example, laws about tax assessment on land, zoning ordinances, and water use and control vary significantly among the 50 states. The major points of local laws must be learned, as well as the process by which the people can change them if the need is critical. For example, among Muslims, many interpret the Koran as forbidding the sale of water, but users may be charged for facilities to transport the water.

The successful expert in an area new to him seeks to avoid ridiculing or violating the religious and other social taboos of the people he is there to help. Otherwise such acts are put down as bad manners, even if done in ignorance.

It has been said that a soil scientist must be a traveller. There is no way to bring a soil into a laboratory any more than one can a volcano or river. He must go to the soils. The soil scientist, of course, does need to collect samples of soil horizons to about 2 meters for examinations in the laboraotry. The same applies to systems of farming. One must go to them. And this means foreign travel, including some in the "bush."

The successful traveller who understands what he sees and who communicates with the people must have good manners and an appreciation of religious systems, social systems, and law. Besides being well informed in his own field of scholarship, he must have an appreciation for the related fields, not as a "generalist" but as an educated specialist. As he learns to help people he may become an operating cultural anthropologist as well.

The useful advisor *must* go to the farms and villages. Usually the capitol city is a poor place to learn specifically about the problems of cultivators. Often the statistics are inadequate or misleading, even in parts of the "developed countries." On field-study trips one must be careful to see representative samples of the soils, the farms, and the people.

Institutions

For the development of agriculture, many institutions are needed. Every country has some balance between those in the private and the public sectors. Postal services and highways are usually public. Railroads, banks for saving and credit, and others may be either. In several countries, cooperatives play a large role in furnishing supplies to cultivators and in marketing their products. Increasingly education is in the public sector. Most advisory (or extension) services for farmers are public.

But however these essential services are supported, they may also need technical assistance to be effective for improving farming areas. Several American scholars have studied this question and issued suggestions for USAID bilateral assistance (CIC-AID, 1968). At least some farm products must be sold that require storage, grading, processing, and shipment. As farming improves, more farm supplies are needed, such as fertilizers, pesticides, and tools.

Honesty, competence, and social vision of the staff are the most important aspects of institutions that service cultivators effectively. There seems to be something in the Scandinavian culture that allows them to have highly efficient cooperatives. Yet I have seen good co-ops run in other countries, as well as some that were run for the advantage of only a few in the village. Here again the route depends on the people. On comparative costs, an experienced expert in economics can be helpful in guiding broad decisions.

Education is vitally important in the general development process. It is helpful to the cultivator to be able to read the directions on tools and machines, the statements on a bag of fertilizer, and so on. Yet we tend to overemphasize both ends of the educational route—the primary and the university—the first for national prestige and the second for individual prestige. In highly status-conscious societies either extreme can lead to inefficiency in the goal of improved living standards. (For a valuable discussion of educational needs for agricultural development see Coombs et al., 1971).

Then too, an educational plan for agricultural development must take account of the growing needs for mechanics, electricians, truck drivers, and the many other skilled workers required. We must continually recall that the unskilled and unproductive people in the village can be almost driven to the big port cities where life will likely be worse (Andreski, 1968, Chapter 3).

A good extension or advisory service is nearly essential where the best opportunities for increased productivity of both the soils and the people require new crops and new practices. Many have almost taken for granted that all good advisory officers have university degrees. This is not always true, especially for those who work in a local area with one or more villages. It may be far better to give a bright young man in the locality who knows the local language and its work habits a 2-year training course beyond the intermediate school. Such a man could get down on his knees and show a cultivator how to fertilize sugarcane (*Saccharum officinarum* L.), for example, or to set out plants from a nursery bed, while one with a university degree might not. Of course, such a young man would need suggestions and some refresher courses from time to time.

A really effective advisory service has the backing of a national cooperative research endeavor, which may encompass several individual research centers that carry on both basic and adaptive research. This second function is commonly cooperative with subject-matter specialists of the advisory service itself (Kellogg and Knapp, 1966). Since *basic research*—as the term implies—looks to the refinement and quantification of older approximations and to the development of new principles, many overlook an enormously important additional function of basic scientists in agriculture. In the process of developing new plant varieties, for example, for high-yield potential, improved quality, and disease resistance, the range of germ plasm is narrowed. They may become prey to some new disease or insect. We can look only to experienced scholars in genetics, plant pathology, and entomology to analyze such a problem. Comparable examples could be given about soils, livestock, and other basic research areas.

Unconsciously, Lewis Carroll summarized the story of modern farming well in his *Through the Looking Glass*. The Red Queen and Alice were running very fast. Alice became exhausted and sat down. She looked around and said, "Why, I do believe we've been under this tree the whole time! Everything is as it was!"

"Of course it is," said the Queen.

"Well in *our* country," said Alice, "You'd generally get somewhere else if you ran very fast for a long time."

"A slow sort of country!" said the Queen. "Now here, you see it takes all the running *you* can do, to keep in the same place. If you want to get somewhere else, you must run at least twice as fast as that"!

The more we increase yields, the more likely is damage from weeds, for example. Our best crop varieties and breeds of livestock are not selected to compete with weeds or to grow well on badly managed soils. The farming systems must be watched and, if something goes wrong, prompt corrective measures need to be taken. Having achieved some progress, care and attention are needed just to hold these advances.

Good living

A basic aim of development is good living on the farms, in the villages, and in the towns. As both the people and the soils become more productive, money will be available. Then useful things must be available to those who earn the money.

I can recall instances where development went faster and better than even the experts had expected. For the first time people had money to spend and yet arrangements had not been made for stores with useful items. Most of the money was wasted on overpriced trinkets. In one extreme example, people had seen pictures in a magazine of electric refrigerators used to store beer. A few bought these and used them as pictured, even though the area had no source for electric current! How much more useful would have been a flashlight or a bicycle!

This does not mean that everything they buy should be practical, nor what either this author or his reader would have recommended. If people have never had any "pretty" things in the home, they are likely to want some, and tastes vary.

At least some education and facilities for spending should be planned so that part of the new income may be used to improve water supplies and sanitation generally. Clever and informed household advisory officers, especially women, have been quite successful along these lines. Most cultivators and their wives, whom I have known, are anxious for their children to live and be healthy. In many societies able sons are the only "social security" they have in old age. They need to know what to do to keep their children healthy.

Most are also anxious for their children to learn to read and to write. They see technical assistance workers like us writing in notebooks and reading papers. Possibly many are so insistent on literacy for their children because they know that literacy helps in politics. And to many of them politics seems to determine who gets what.

2

Soils for many uses

The total area of arable soils in the world being used for field and horticultural crops approximates 1.4 billion hectares, or 3.5 billion acres, out of a total potential of 3.2 billion hectares, or 7.8 billion acres. The estimates of soil areas were made on the assumption of using only known supplies of water, without special desalting industries powered by oil or nuclear energy to increase irrigation (Kellogg and Orvedal, 1969). (See maps in Figures 8 and 9).

A considerable portion of the unused potential of arable soils—1.6 billion hectares—lies in areas now lacking essential transport and other infrastructure for commercial farming. Yet road construction is now going ahead in the Amazon Valley and in Central Africa.

In this chapter we deal largely with soils, the natural support for most plant and animal life. Since some confuse the words *soil* and *land* it may be well to define them. *Soil* is a collective term for an enormous number of individual soils and kinds of soil. *Land* is commonly thought of as real estate. The classification and value of a tract of land depends on its location, size, distance from market, potential use for buildings, and other factors besides the potential productivity of the soil.

A classification of soils is one dealing with them as natural bodies, regardless of location and social facilities. A classification of land deals with tracts having unique boundaries defined by latitude and longitude or in some other official grid system of a cadastral survey that is specific for each tract.

Accurate soil maps are widely used in land classification, along with other essential data. Predictions of the most likely outcome of alternative uses of the kinds of soil are given in most detailed soil surveys. But these are not "recommendations" for the

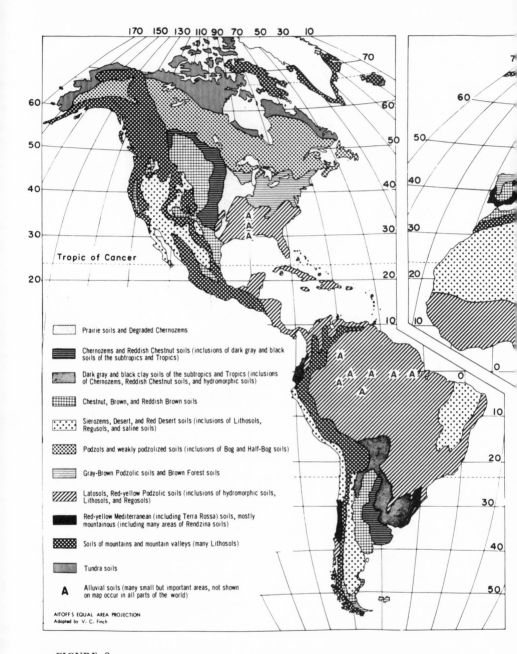

FIGURE 8

A small-scale soil map of the world based on the system of soil classification current in the United States from about 1950 to 1965. For detailed descriptions of the map units, see Kellogg and Orvedal (1969).

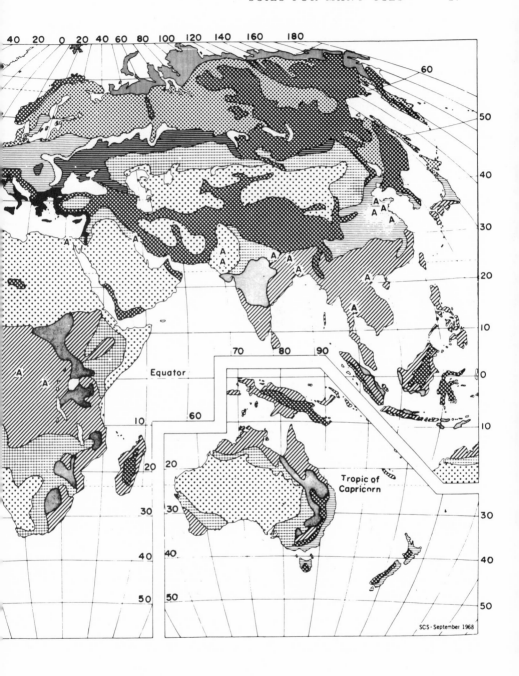

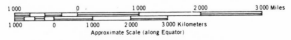

U. S. DEPARTMENT OF AGRICULTURE

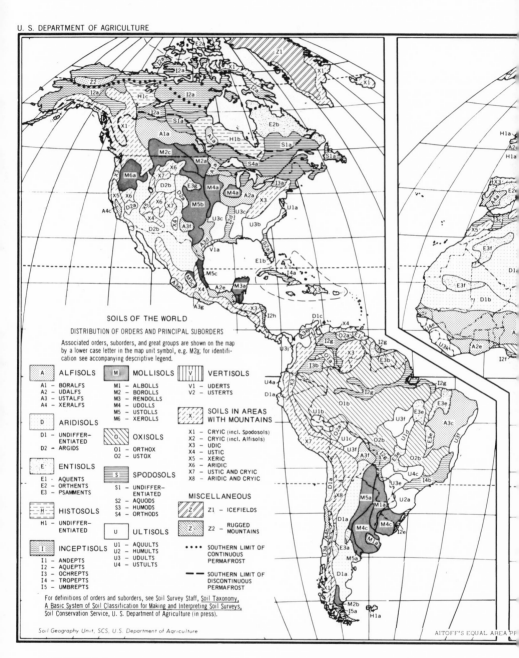

FIGURE 9

A small-scale soil map of the world based on the current system of soil taxonomy in the
United States. For detailed descriptions of soils, see Soil Survey Staff (1975). Courtesy
Soil Survey, SCS, USDA.

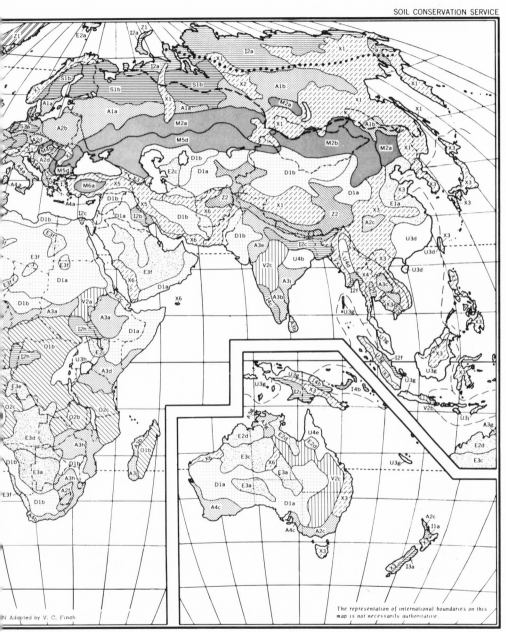

N Adapted by V. C. Finch

The representation of international boundaries on this map is not necessarily authoritative.

MAY 1972
USDA SCS HYATTSVILLE MD 1972

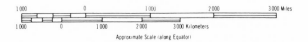

Approximate Scale (along Equator)

use of specific tracts of land because of the other features of land, markets, transport, and the skills of the operators.

The world has a wide variety of kinds of soil, with associated climates. With human labor they furnish most of our food, textiles, timber, and paper as well as other raw materials for industry. They are used for parks, wild-life preserves, lawns, and play areas. Besides supporting plants, soils serve as foundations for homes, highways, airports, and many other structures. Some kinds of soil highly useful for plants cannot support structures except with costly modification of the soil.

Actually the way people live and the kinds of societies they have developed depend a great deal on the kinds of soil they have and the skills they have learned to accomodate to them. That is, soils perform well or badly, depending on how well people select uses for them or, to turn the problem around, how well they have selected kinds of soil to use, and how well they have designed systems of management, or manipulation, in accord with the unique combinations of properties of the soils. These determine their potentials to respond to *specific* uses.

THE NATURE OF SOIL

Early soil study in Western Europe was mostly in the laboratory and on the experimental plot. Soils were looked at mainly as geological fine earth to receive manures and fertilizers. The dominant idea in the middle of the nineteenth century has been called the "balance sheet theory of soil productivity." One had to put back what he took out. Soils in Western Europe were used mainly in small farms and each family knew the potentials of their own soils fairly well.

Old Russia on the other hand was an enormous empire with large areas of highly contrasting plant associations, climates, and landforms. The main questions were what soil resources did the country have and how could they be used.

About 1870 the great Russian scholar, V. V. Dokuchaev formulated the outlines for the basic concept of the nature of soils and their formation now used throughout most of the world. He found that soils had geography. They did not occur promiscuously over the earth, but appeared in association with specific climatic belts and vegetative zones and, locally, in relation to relief, land-form, and geological materials.

The number of kinds of soil must be of the order of 500,000 to 1 million. As results of research and experience accumulate from the least well-investigated parts of the world, the numbers of recognized and named kinds of soil will doubtless grow.

The word *soil* is a collective term for an enormous number of individual soils, comparable to individual animals, plants, and other natural objects. Each of these individual soils occupies an individual piece of landscape. Each has depth, shape, and area. If we make a deep pit in any one we find that the soil is made up of layers or *horizons*.

The collection of all the horizons of a soil as exposed in a pit down to the nonsoil beneath, which may be hard rock or fine earthy material, to about 2 meters, is called the profile of the soil (Figures 10, 11, and 12). Its description includes precise designations of the color, texture, structure, thickness, and so on of each horizon, supplemented by laboratory examinations of the physical, chemical, and biological properties.

FIGURE 10

Profile of Podzol (Spodosol) in Maine showing organic mat, light gray leached A2 horizon, and brown B horizon over glacial till.

Not all natural soils have smooth surfaces. Several natural processes can cause either the surface soil, or horizons beneath, or both, to have obvious microrelief, or waviness. As trees blow over, especially those with shallow root systems, roots and soil come up to form mounds and swales, called *cradle knolls* (Figure 21). Burrowing animals also cause unevenness, as can heavy freezing, extreme shrinking and swelling of active clays during alternate dry and wet seasons, and several other natural processes.

Since the soil profile has only two primary dimensions, while a soil individual has three, the profile description alone does not suffice for close study. Enough descriptions are needed of the thickness and depth of horizons to include the seasonal cyclic variations, say about 1 to 10 square meters.

Several kinds of soil have such "cyclic" variations. In the dry season the Vertisols, or Tropical Black Clays in the older nomenclature, mulch at the surface and form deep cracks. When the rains come the surface mulch goes into the cracks. Then when the soil is all moist something must give way. So the surface is pushed up between the cracks, hence the new name, Vertisols, or "turning soils." This leads to a microrelief called *gilgai* (Figures 13 and 14).

FIGURE 11

Nearly black Chernozem (Mollisol) in North Dakota, with a horizon rich in calcium carbonate beneath. These are excellent soils for wheat. Courtesy W. M. Johnson.

FIGURE 12

A leached acid soil in New Zealand with a fragipan—a compact layer (Fragochrept). These soils are only moderately productive.

FIGURE 13

The surface of a Vertisol
(Ghana) in the beginning of
the dry season. Note the
surface mulch and cracks.

FIGURE 14

Gilgai microrelief in a Vertisol (Ghana) is pictured here.

The lowest sampling unit with a full description is now called a *pedon*; and a soil individual is thus also a *polypedon*.

In their physical, chemical, and biological properties, the horizons of a soil may contrast with one another only a little or very greatly. The temperature and moisture of the horizons also may vary greatly or only a little during the year. Then too, total depth of soil, stoniness, and slope of unlike soil individuals may vary slightly or greatly from one another.

Since each soil is a three-dimensional individual, its boundaries with different soils or with nonsoil, such as bare rock, can be plotted on a map of suitable scale. The area of an individual soil can vary greatly. Thus we see many large fields with nearly uniform stands of crops or trees. Other small areas of only a few hectares may have three or four highly contrasting kinds of soil within them.

All of the soils that have the same combinations of properties can be grouped together into a *kind of soil*. Each individual soil belonging to a kind has a name and a unique combination of soil characteristics, including seasonal regimes of moisture and temperature. Each soil of a kind has developed under a unique combination of soil-forming factors. These include the active factors

FIGURE 15

Much of the improved pasture was due to the introduction of earthworms.

FIGURE 16

An overcropped, much fired area of Oxisol (Earthy Red Latosol—Kellogg and Davol, 1949) near Mt. Hawa (which has a dry season) in the Republic of Zaire (Belgian Congo). The cover is a bad weed grass—*Imperata cylindrica* (Cogongrass)—which can be eradicated only by deep machine plowing. When the soil is covered with this grass, and many millions of hectares in Africa are, the soil cannot be used for farming or grazing by most local people using only hand tools.

of (1) climate and (2) living organisms, including plants, animals, and microorganisms acting on (3) the earthy parent material as conditioned by (4) the relief of the landscape and by (5) the age of the landform, especially of its surface. Many soils are very young and some are very old.

Soils that have had their horizons significantly altered under use are separated from others. Examples include soils that have had more than normal erosion or soil blowing that have caused removal of upper horizons. Others have been thickened by accumulations of sediment. Many soils have been changed by grading for construction. Some soils in long-cultivated areas have highly thickened surface layers, commonly rich in plant nutrients and organic matter, due to long periods of heavy manuring or composting. With these practices, organisms in the soil may be changed, including the small burrowing animals (Figure 15). Others have received harmful accumulations of salts from older irrigation either improperly designed or spoiled by destructive wars.

Curiously, this basic concept of the nature of a kind of soil and its related landscape and living matter first developed by Dokuchaev about 1870, was "rediscovered" around 40 years ago and called an *ecosystem*. Thus, where both concepts are defined *precisely* in the natural landscape, the kind of soil and the ecosystem, or potential ecosystem, are closely related.

Yet ecosystems are not easily named, partly because the soils are rarely examined. The current vegetation may have followed older clearing or fire (Figure 16). A large part of the soils of Africa, for example, now have poor savanna vegetation as the result of thousands of unrecorded "accidents" of clearing and fire (Whyte, 1962). Potentially good arable soils and others clearly unsuitable for farming with current practices can have either different or similar savanna vegetation now.

Most kinds of soil can be identified in some recognized system of soil taxonomy. Because of the generally excellent relationships between the many soil research institutes of the world, and frequent international meetings of soil scientists, most know one another's systems. Gradually the scientific nomenclature throughout the world is becoming increasingly similar, and more soil scientists in the several countries understand each other's terms and soil names.

Without having the soils described and named in some standard taxonomic system it is not possible to quantify either their characteristics, which determine their responses to management, or the predicted yields of alternative crops and trees under specified management.

The individual kinds of soil can be grouped into more broadly defined units as in other systems of taxonomy for natural objects. With soil maps that show the kinds of soil in a standard system of taxonomy, the results of research and experience can be transferred safely from one local area, country, or continent to another. The multitude of heartbreaking and costly failures from simple "trial and error" can be avoided or greatly reduced.

As more research has been done on the properties and behavior of soils, the taxonomy has been improved and especially the interpretations of the responses of kinds of soil to management and manipulation. Quantitative data and predictions of alternative uses for kinds of soil are essential for use of soil maps in agricultural development, whether for improving old areas or for developing new ones.

Interpretations of the kinds of soil can be given in text descriptions; but for easy use they are given in tables. For example, a yield-and-practice table gives the estimated yields of each adapted crop for each kind of soil in a survey area with one, two, or three physically defined, alternative systems of management. These are invaluable for operational planning of a farm.

For some purposes, such as land reform and especially the settlement of new areas, it is also helpful, indeed nearly essential, to give a productivity index based on the best yield of each adapted crop for the area as 100. For the principal crops to be grown practically, these productivity ratings can be averaged into a more general productivity rating for several field, horticultural, or industrial crops. For determining adequate allotment size, one cannot use total hectares alone since the kinds of soil vary widely in yield potential. Thus allotments of 10, 20, 30, 40, or some other arbitrary number of total hectares can be very unjust. By using the productivity index each allotment can be adjusted in size to have approximately the same potential productivity.

Following the same principle, the forest site indices of adapted trees are given by kinds of soil. Similarly, for extensive grazing, yields of useful forage can be given for each kind of soil under excellent, good, fair, and poor range conditions, each of which also needs to be defined. The yield variations under contrasting seasons, such as unusually dry, average, or unusually moist, need to be given.

In some countries all the kinds of soils are grouped into three to eight classes, commonly called "capability classes" for extremely broad interpretations according to their general limitations and potentials. These are broad indeed because some kinds of soil suitable for horticultural, floricultural, and other special crops are quite unsuited to the ordinary field crops, and the other way around. Such general groupings allow only rough preliminary appraisals of broad areas but lack the essential specificity for actual operational planning on farms by either the cultivator himself or by him with the guidance of advisory people who may be assisting him.

Yet some broad groupings of soils into associations can be helpful as a general guide to advisory people for trial. The use of the catena[1] grouping is a good example. A map showing the differ-

1. A soil *catena* is a grouping of kinds of soil within a limited climatic region formed from one kind of parent material but differing in characteristics owing to differences in relief or drainage.

ent catenas can be made at fairly small scales and the kinds of soil recognized within each of them. Such a scheme is outlined for broad use in Buganda, a province within Uganda by S. A. Radwanski (1971). With good aerial photographs such maps can be made with less cost than a detailed soil survey for operational planning. In this example he gives alternative potential uses for broad groups of soils.

Civil engineers found in the 1920's that the soil classification used for soil surveys, originally made for planning farming and forestry, could be very useful in the location and especially the design of highways. The high cost of failures in cement-concrete roads greatly stimulated the joint use of surveys by soil scientists and engineers. From the early trials this work has been greatly extended to other kinds of roads, airports, to both town and country planning, and for location of buildings and treatment of soil surfaces. (As examples among many, see Michigan Department of State Highways, 1970; and Robnett and Thompson, 1970).

Now a wide variety of soil engineering data and interpretation are also given in tables for easy use for each kind of soil. Such tables suggest the stability of the kinds of soil for foundations of

FIGURE 17

A so-called Kauri Podzol in the northern part of New Zealand. Kauri trees (*Agathis australis*) shed their leaves and bark which contained strong reducing compounds. These compounds reduced the iron, which moved from the surface to form a hard dark brown horizon beneath. Other kinds of trees have this influence, but few, if any, to such an extreme degree.

highways, buildings, and other structures; they suggest permeability to water, usefulness for subgrade, usual variations in depth to rock, and other qualities of the soil relevant in the survey area. For example, in places being built up with houses, streets, and other structures that cause reduced permeability of the soils to water, estimates by soil scientists and hydrologists, working together, are needed to predict potentially critical flood hazards. Obviously such studies are needed in advance of construction. If done only after construction they merely explain the great troubles and losses.

DYNAMICS OF SOILS AND LANDSCAPES

As we saw in some detail in the explanation of shifting cultivation, green plants, powered by sunlight, manufacture their own food from the plant nutrients taken from the air, water, and soil. Initially the very young soil may start as little more than fine earth and small organisms.

The plants remove the mineral nutrients and the nitrogen from the soil, although some nitrogen is captured by rain falling through smoke and especially that fixed during thunderstorms. Some mineral nutrients are taken in from dust on the leaves (Figure 1).

Obviously plants and microorganisms have an enormous influence on the soil. Some of the small organisms fix nitrogen from the air. If plants take in large amounts of calcium from a deep root system, or from dust on the leaves, the upper soil layers may be considerably enriched in calcium. In fact, the odum tree (*Chlorophora excelsa*) of the secondary tropical rain forests of parts of West and Central Africa takes in so much calcium that huge masses of calcium carbonate may form in old wounds. This explains how some local people have lime for whitewashing walls where there is no other local limestone to be had. Yet a tree growing nearby may take in much sulphur and under it the soils may be acid.

Other trees drop organic matter as leaves and bark that decomposes to give an organic solution of high reducing powers. Under these trees the iron is readily leached from the surface soil (Figure 17).

FIGURE 18

A braided glacial stream in New Zealand. It is a source of loess when the weather turns dry and windy.

FIGURE 19

The braided Matanuska River in Alaska with the fine sediments blowing out of the dry river bed as loess. (The photograph appears hazy because of the heavy dust blowing at the camera.)

All of the five factors that make a soil of a certain kind are dynamic. Many parts of the world have had tropical climates, including northern Alaska, Canada, and the USSR, else one could hardly account for the petroleum deposits.

During the Pleistocene period, beginning roughly 3 million years ago and ending roughly 10,000 years ago, the world had various intensities of continental glaciation, which ground rocks and moved the particles of earth, all the way from clay to huge boulders.

Some of the most responsive soils in the world have developed from deposits of fine, silty material blown out of great river valleys from droughty landscapes, and even from areas which are now sea, when the water level in the oceans was much lower, and deposited in great blankets or even dunes. Many of the soils in central United States, for example, have developed from blankets of silty material blown out of the braided stream beds during the Pleistocene. The same is true of large areas of China, and the northern parts of western and central Europe and much of the great "fertile triangle" that begins in eastern Europe and extends east far into the central part of the Soviet Union (Figures 18 and 19).

Even many of the old soils in the eastern part of the United States have silty surfaces resulting from violent dust storms during long periods of extreme drought in the Great Plains. Such droughts can be expected every 20 years, but their duration and the area covered varies. Such a period occurred during the early 1930's in the United States. It was worsened by cultivation of unstable soils, but it resulted mainly from several years of extreme drought. These storms had gone on for many centuries before there were any plows in North America.

A high proportion of the food crops of the world is also grown on soils formed from both old and young alluvial deposits along great rivers and small streams, and in their deltas. Much of this earthy alluvium came originally from natural erosion, from both slips and gullies in the hills, and from "cutbanks" along the rivers. The waters from fertile mountains brought organic matter along with the earthy material from slopes that were covered with trees and other plants. It was this kind of sediment in the alluvium that made the soils of the old Nile Valley so productive.

Coastal areas near the oceans have been altered and realtered a great deal by the wave actions of the sea, by accumulation of sediment from streams, and by salt water from the sea carried onto the soils by typhoons and hurricanes. Some coastal lines have

FIGURE 20

This is a typical small mud flow in New Zealand.

steep cliffs while others have low, gently sloping plains. Within these low areas, at slightly lower elevations, are irregular patches of swamps and marshes. Shorelines are also complicated by the fact that over the millions of years the oceans have been both higher and lower than now.

As weathering proceeds on slopes, fine and coarse material is moved downward by the pull of gravity assisted by water (Figure 20), frostheaving, and the windthrow of trees (Figure 21). On very steep slopes the weight of the growing trees themselves result in landslides. Much of the weathered material on the slopes of mountains is unstable. The construction of a road or even a skid-trail in a critical place may concentrate the water just enough to initiate a landslide.

Then, of course, volcanoes throw out new material over the soils, sometimes with dramatic suddenness. A wide assortment of material comes out of volcanoes. Some of it is excellent for soil development, but other materials are not. Heavy rains may move it far down the slope. The very fine ash may be carried great distances by the upper air currents. This is also true of the dust from soil blowing in deserts and during very dry periods in subhumid climates.

Thus, landscapes are being changed continually through these kinds of natural processes. They go on today. Old landscapes are being torn down and new landscapes created. For some of these changes men are almost helpless. Other drastic changes can be considerably avoided through fire control, grazing control, and by avoiding plowing or road building where the balance is delicate.

Some people appear to take the absurd view that all such movement of earthy material can be controlled by man. Many create the impression that farming somehow injures the soil and the environment. Generally, however, farming has improved the environment, although there are exceptions where unresponsive soils have been cultivated and where even good soils were badly managed.

On individual fields and gardens it is highly important to control runoff and to protect soil from strong winds, especially during dry periods. Yet we must recall that many of the most productive soils originated on landscapes resulting from the accumulation of alluvium and windblown silt.

Some great fires were set accidentally, of course, but many started from lightning and from the sparks made by great stones crashing down the mountain in a dry period. After settlers built roads in the American Great Plains, the great "prairie" fires ceased. In many parts of the world fire has been a great enemy of productive soils. Perhaps the best examples can be drawn from Africa

FIGURE 21

Here is a typical "wind throw," which causes "cradle knolls." The tree roots bring up soil. When the roots rot, it results in a mound near a pit. Such cradle knolls are common on ill-drained soils where tree roots are shallow and strong winds blow occasionally.

(Figure 16). We know that during much of the Pleistocene period even the Sahara was covered with vegetation. At great depths in parts of the Sahara are huge deposits of ground water—fossil water, not being recharged now. After the glaciers receded the Sahara became dry again. Many areas south of the Sahara have long dry seasons. It is doubtful that Africa had much "natural" savanna a few thousand years ago, except near the sea and in high mountains. Trees were cut and burned to grow food crops, and many areas were burned to encourage grasses for pasture. Few of the early people left records of any kind, except in the soils.

Because of the great changes in temperature, rainfall, and earth movements, soil formation may take one direction for a few hundred or thousand years and later change toward other kinds of soils. It is a brave soil scientist indeed who *knows* from field examination alone that the surface soil he looks at was developed from exactly the same kind of earthy material that lies directly underneath, or even that the lower horizons in the soil profile were developed under the same climate as the surface ones or the still deeper ones.

A growing literature is accumulating on the relations between the old soils, called "paleosols," buried in more recent deposits, and the associated geomorphic changes in landform. As two examples among many, the reader may look at an early study by Robert V. Ruhe in the eastern part of Zaire (1956) and a later one in southwestern Iowa (1967). Several examples are also given in Schultz and Frye (1968) as related to coverings of loess and other eolian deposits of the world.

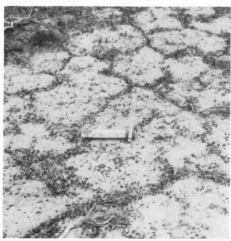

FIGURE 22

Natural frost cracks and pebble rings on undisturbed soil are seen here in Idaho.

Still a few soils have been isolated from both animals and people for a long time. For example, a lava flow in Idaho split into two streams that coalesced a bit further down the gentle slope. The cracked lava could not be traversed by grazing animals. So the soil was undisturbed for many thousands of years. In winter it froze and in summer became warm. Figure 22 shows a nearly micro-tundra effect with pebble rings in the cracks.

Many scientists have studied the mutual relationships of plants and soils under changing environments. As one good example the reader may look at J. E. Weaver's (1968) summary of his life's work.

SOIL FORMATION AND CLASSIFICATION

A detailed explanation of the vast literature on these subjects would require many volumes. The basic principles have been enormously clarified and elaborated as more and more soils have been studied in both the field and the laboratory.

To have useful soil maps a classification of the kinds of soil is essential. In the beginning, soil scientists (or pedologists) had the field methods of geological exploration and correlation and the laboratory methods of chemistry and physics, to which were later added those of soil mineralogy and biology.

At the insistence of Henry A. Wallace, then Secretary of Agriculture, the 1938 Yearbook of the Department of Agriculture, *Soils and Men,* was devoted to the nature, use, and conservation of soils. For that book a considerably improved system of soil classification (taxonomy) was developed. (The principles were explained simply for the general reader by Kellogg, 1941).

Both at that time and since, many have asked me—"when are you soil scientists going to stop changing the names, descriptions, and interpretations of soils"? The nearly obvious answer is, and will continue to be—"when soil scientists stop learning more about soils and when people stop having new problems, arising from new crops, new varieties, new machines, new types of construction, new sources of pollution, increasing demands for space and products from the soil, and so on."

As in other sciences, many soil scientists tend to become a bit overspecialized. Some have devoted their attention almost exclusively to the so-called "basic" scientific principles of formation and classification of soils with little thought to their potential uses and response to management. Since the work requires funds, the lack of benefits to people results in low support for it. This has been a great pity in some countries.

On the other hand, a few soil scientists have given such exclusive attention to the "practical" side that they did not use the opportunities for scientific research and their work failed because of both inaccurate soil maps and predictions about soil use and management. To produce a good soil survey for people to use requires excellent scientific work and full interpretations of the soils for their expected alternative uses in quantitative terms.

Thus a system of soil taxonomy must be tested through research and against its usefulness in predicting the results of use and management. In the United States and elsewhere during the 1940's the system outlined in the 1938 Yearbook, and other systems, were severely tested. Weaknesses in specificity were found. Some individual soils, given the same names, turned out not to be similar in their responses to use and management. Properties had been overlooked that influenced the significance of other properties. Too much emphasis had been placed on their surface horizons that were changed by normal farm use. Neither the temperature nor the moisture in soil horizons during the year were formerly considered as soil properties. This resulted in wide differences in the use-potential, especially of young soils with only weakly developed horizons that looked alike, or among old soils that were developed under quite different climates than their present ones. One can by no means assume that all the important properties can be preserved in samples of soil horizons put into bags for the laboratory. For stoniness, for shape and degree of slope, for greater than normal losses or additions by soil erosion or blowing, and for other external features, measurements must be made in the field.

All the properties of a soil, including seasonal moisture and temperature, must be considered *together* to arrive at a sound judgment for its classification or interpretation because each one interacts with the others.

In the United States in 1951, the decision was made to develop a new system to account for all the soil characteristics as the soils exist, both those that had been modified by use and the strictly natural ones. Fortunately soil scientists from several other countries were glad to join in this effort and to help test new proposals as well as to add their own.

Since it would require many years to develop a final text, the early attempts were called "approximations," beginning with the "First" in 1951. In 1960 on the occasion of the 7th Congress of the International Society of Soil Science in Madison, Wisconsin, the 7th Approximation was published (Soil Survey Staff, 1960). Supplements were issued in 1964, 1966, and 1967.

It is now expected that the full revised Taxonomy will be published sometime in 1975, along with a third edition of the *Soil Survey Manual* (second edition: Soil Survey Staff, 1951).

Guy D. Smith, who had staff leadership in the National Co-operative Soil Survey of the United States for this effort, gave a series of lectures on the system at Ghent University that were published in *Pedologie* (Smith, 1965). Since the scheme has now been used in essentially all soil surveys published in the United States after about 1966, several examples of its use are available in libraries. At the same time soil research institutes in other countries have been revising their systems. For example, the French system is outlined in a new third edition by Duchaufour, 1970.

By studying the soil maps given in Figures 8 and 9, the distribution of the broad groups of kinds of soil in the world can be seen in both the older system (1938) used in the United States and of the newer one now in press. (For unfamiliar terms see Glossary.)

Beginning in 1961, with collaboration from UNESCO, the Food and Agriculture Organization of the United Nations (FAO) has had the participation of many countries toward the development of a small-scale $(1:5,000,000)^2$ soil map of the world with a "unified" legend. Although few may agree with all names and aspects of the classification, the effort has brought together many soil scientists so that they are better informed about the soils of the world and the research results available. In the process FAO has published some 40 reports on conferences and field excursions. The first maps were published in 1971. These were for South America (FAO-UNESCO, 1971).

Soils of the tropics

Since so many of the less developed countries are tropical perhaps a brief general statement may be useful here. In the tropics there

2. Fractional scales are easily translated regardless of the system of measurement. A scale of 1:5,000,000 means unit, inch, or centimeter, for example, on the map to 5,000,000 of the same unit on the ground.

are very young soils from recent alluvium and volcanic ash and many with fair amounts of fresh minerals containing plant nutrients. The old stable landscapes of the humid tropics that receive little alluvium, dust, or volcanic ash, and that have been under considerable leaching during the rainy seasons, are generally low in plant nutrients. This applies especially to those that may have been cleared for farming or for other reasons many years ago, and have been invaded by savanna grasses that are usually fired in the dry season. Here the nutrient cycle from soil to plants and back again is interrupted since the ashes may wash or blow away. A high proportion of the upland soils are red, reddish brown, reddish yellow, or yellow. Their classification and nomenclature has been going through many evolutions during the past 30 to 40 years.

Many have been called *lateritic soils* because some do have laterite beneath. They have also been called *red loams* because many are reddish and because the upper soil horizons are relatively porous and nonsticky for the amount of clay in them. The term *ferrallitic* is also used because of the reddish colors suggesting iron oxide.

In an early study of these soils I used the coined name "Latosol" for them with several subdivisions (Kellogg and Davol, 1949). More recently the term *Oxisols* is favored for those where most of the soil material is well oxidized and porous. Associated with the Oxisols are sloping soils from basic rocks, such as basalt, that are relatively rich in mineral plant nutrients, which were formerly called *Reddish-brown Latosols*, among other names. Now in the new U.S. Soil Taxonomy they are called *Tropepts*, suggesting youthful tropical soils.

Also in the tropics and subtropical areas are many other soils common in warm-temperate regions. We have already mentioned the Vertisols (for "turning soils") formerly called "Tropical Black Clays" or "Black Cotton Soils" in the older literature. These soils also are found in warm-temperature regions. They are formed mainly in places with wet and dry seasons from fine clay deposits.

These soils are difficult to till with hand tools. Interest in their use for farming increased greatly as heavy power machines became available for plowing and for smoothing the humps and hollows of the gilgai microrelief.

Since an enormous literature is already available to the interested reader I see no need to go further with this subject except for a few special soil conditions that have been commonly misunderstood by some writers.

Laterite

Curiously, since about 1964 the hazard of laterite[3], which does appear in some tropical soils, and in paleosols now under different climates, has been enormously exaggerated by a few researchers. It is an interesting phenomenon of some local importance here and there in the tropics and even as relics in present temperate regions.

The areas of soil with laterite are commonly small. As the reader looks at the maps in Figures 8 and 9, he sees that the tropical areas, as a whole, have other kinds of soil, many of which are highly responsive to management for farming.

Laterite (or plinthite) is formed over a long time in or beneath soils on very old, smooth, stable landscapes in the tropics that have pronounced wet-dry seasons and water tables that fluctuate from near the surface in the wet season down to considerable depths in the dry season. The soil material is thus leached of its bases, such as calcium and magnesium, and then of its fine silica. Iron oxide and alumina accumulate from the losses of the other elements. In many areas with laterite, additional iron and alumina has been added from the water table by accumulations from seepage out of adjacent higher landscapes (Alexander and Cady, 1967). (See Figures 23, 24, and 25).

So long as the surface layers remain intact, the lateritic material remains fairly soft or doughy. Those made up of nearly pure alumina are good sources of bauxite for making aluminum. Those very rich in iron oxide can be used for smelting iron.

Once this doughy laterite is exposed to wetting and drying it hardens irreversibly to rock. This happens after geological uplift and a new erosion cycle and from the deepening of a stream and its tributaries. If the vegetation is removed and accelerated soil erosion is permitted to expose the material, it hardens to rock called laterite crust. Curiously, the people of Kerala State of southern India learned over two thousand years ago that they could farm soils with doughy laterite beneath *provided* they never plowed the soil nor exposed it to sun and rain. They grow a wide variety of herbaceous and woody crops in mixed culture with all harvesting and planting by hand.

3. *Laterite* is an old term, presumably derived from the Latin word *later* suggesting *brick* because some people use it for bricks. Later the words *laterite* and *lateritic* were applied to many tropical soils both with and without "laterite." So the word *plinthite* was coined to substitute for it in the new classification (Soil Survey Staff, 1960). But here I shall use the old term.

FIGURE 23

This is a profile with doughy laterite (plinthite) about 50 cm below the surface. The lower material is commonly called the "mottled zone." Near Lubumbashi, Zaire.

FIGURE 24

This is the associated landscape of Figure 23 with laterite beneath and termite mo above. If the tops of these mounds are smoothed, vegetables can be grown on the

FIGURE 25

A large tree pushed a few roots through cracks in the laterite layer. As the roots grew they forced boulders of laterite to the surface. South of Perth in Australia.

Many people in the tropical areas that have doughy laterite uncover it and cut it into bricks, usually about 75 to 100 cm long, about 30 cm wide and roughly 15 cm thick (Figure 26). These become very hard after direct exposure to sun and rain for a few months. Some even cut out circular well curbings and statues.

In places where the total iron and alumina in the material is too low for doughy laterite, concretions only may form. In a later cycle of soil formation, in the tropics, where such soils are on low-lying terraces they may receive additional alumina and iron from seepage and be hardened into "pisolithic laterite."

The natural exposures and crusts of laterite are dramatic. I have even seen beautiful specimens under geologically younger sandstone near Lincoln, Nebraska, and in other parts of North America. Examples can be seen under the soils of the Sahara where it could not form now. Specimens may be seen along the Malabar Coast of India, near Calcutta, in Indochina, in Brazil, in much of Africa and Australia, and elsewhere. But the point I want to emphasize is that its total area and importance have been greatly exaggerated.

With a good soil survey in advance of land development, which is essential anyway, soils with laterite near the surface are easily avoided as well as much larger areas of other kinds of unresponsive soils. Notions that great areas of soils of the Zaire and Amazon river basins are very hazardous for farming because of the potential of laterite crust are absurd. But, of course, some areas do exist in both places. In special situations with a little additional surface soil, some soils with laterite are used (Figures 27 and 28).

Acid sulphate soils or "cat clays"

Since these special soils are not so dramatic as those with potential laterite, they are even more likely to cause trouble in development without soil surveys. The soils are not limited to tropical and subtropical areas. They do not represent a widespread problem but are mentioned because of the potential for some serious local mistakes from which a few people near the sea have suffered costly failures. Now they are easily recognized by most soil scientists.

Most, but not all, of these are small areas fairly near the sea in low-lying coastal plains or deltas where sediments added by the streams are from nonlimy materials. These are wet soils that may be largely or partly organic. Hurricanes and typhoons from the sea bring in salty water with sulphate. Under wet conditions with abundant organic matter, the sulphates are reduced to sulphides

FIGURE 26

An Indian cultivator is cutting doughy laterite (plinthite) into bricks. Kerala State, India.

FIGURE 27

Cashew nuts are grow-
ing with laterite a few
centimeters below the
surface. West Bengal,
India.

FIGURE 28

Hand-made terraces appear in West Bengal with laterite some 30 cm below.
The pattern permits an inspector to check the amount of earth moved.

FIGURE 29

Typical landscape of rushes on acid sulphate clays in the Mekong Delta near Saigon, Vietnam.

FIGURE 30

This cut of a bit more than 1 meter is in a clayey alluvial soil with acid sulphate clays in the lower one-half. These can be used for growing rice. Mekong Delta near Saigon in Vietnam.

and iron pyrite. So long as the soil remains wet, nothing seems to happen. The surface 5 to 10 cm may even have fresh water from streams and support "fresh-water" plants.

But if the soil is drained, the sulphides oxidize to give sulphuric acid; the soils become extremely acid and have highly toxic soluble aluminum, with pH values as low as 2 or even less in a few extreme instances. The ditch banks have large mottles of white and brown after drying.

The very acid soils, certainly those below pH 3, are difficult to reclaim because of the enormous amounts of lime that would be required. And such amounts would raise difficult problems with secondary plant nutrients. Where the mistake has been made, wildlife areas can be restored only by routing the fresh water over the soils.

Some acid sulphate clays are used for rice by keeping the soil wet most of the time except a short period during harvest, as was done in the old days of rice production south of Charleston, South Carolina. And I have seen rice growing on these soils near Cochin, India, and elsewhere.

These small areas of acid sulphate soils are a hazard for both farming and construction in many low-lying areas near the sea in West Africa, Indochina (especially in the delta of the Mekong River), Malaya, India, northeastern South America, North America, and elsewhere (Figures 29 and 30). Any attempt to use such areas should be preceded with detailed soil surveys backed up by a good laboratory (Edelman and Van Stavern, 1958; Moormann et al., 1961; and Dost, 1973).

Extremely acid soils can also be expected in other places where pyrite is near the surface or where water that has leached deposits containing it enter the soil.

Organic soils

Many soils are developed almost wholly from peat. With these the earthy parent material consists of a wide assortment of organic remains, including the feces of aquatic animals as well as the remains of plants, along with some earthy material washed in during floods, blown in, or dropped as volcanic ash or cinders. These soils are especially prominent in geologically young areas along low-lying ocean shores, in large river deltas, and especially in regions covered by the last continental glaciation. Enormous numbers of small lakes have been filled with peat, in parts of northern United States and in Canada, for example. Many of the larger lakes can be expected to fill in future years.

In old settled areas, sediment from streams has covered a part of these organic deposits. Such places are very hazardous for construction of buildings and highways. If drained and cultivated, some of the organic soils are highly productive (Figure 31). Yet in temperate areas, and especially in the subtropical and tropical ones, the organic matter can be expected to decompose more rapidly than it can be restored economically. This does not mean that the soils should not be used but most, especially in warm countries, will gradually settle and finally disappear under intensive use for cultivated crops (Skoropanov, 1961). In areas with nearly continuous rain, snow, or mist, the peat grows over the hills. These peats are called "climatic moors" (Figure 32).

Salty soils

In developing soils for farming an excess of salts can be a serious hazard. A few salty soils may be expected near the sea, where the salts have been brought in by hurricanes or typhoons. Yet in Australia salts from the sea are carried far inland by the west winds, by leaps from one salty area to another. They are also found in closed dry basins.

Mostly salty soils are expected in arid, semi-arid, and sub-humid regions (Figure 33). One may say roughly that salty soils

FIGURE 31

This excellent soil from peat is found near Minsk, USSR.

FIGURE 32

Sheep are grazing on hill peat ("climatic moor") in Scotland.

FIGURE 33

Somewhat salty desertic soil is found in Nevada with creosotebush (*Larrea* L.)

occupy places in these regions analogous to the places having wet or organic soils in humid regions.

The salts may be distributed throughout a potential rooting zone, concentrated in or near the surface, or in layers beneath the surface. Many kinds of salt can be present, but generally sodium chloride is most common. Much of it arises from water seeping through rocks, or soils at higher levels. Then wherever the water evaporates, the salts are left. Old dried lake bottoms, or playas, are commonly very salty. But of course these salty places may be again covered with sediment or wind-blown silt or sand.

Some of the saline soils are simply charged with salt. The sodium ion is absorbed on the fine particles. Then with increased leaching, naturally or artificially, the excess salt may decrease and the sodium comes back into the soil solution as alkaline sodium hydroxide. In the presence of carbon dioxide, the result can be sodium carbonate and a highly alkaline soil. The organic matter in the soil is dissolved and coats the particles so that the soils are commonly black. In the old days, people spoke of "white alkali" soils and "black alkali" soils. Now they are defined by the amount of salt present and the amount of exchangeable sodium (Richards, 1954).

These accumulations of salts, and their effects make some soils highly alkaline, and are especially critical in planning soil use with irrigation. Crops vary in their tolerance of salts and high alkalinity. Cotton and especially date palms (*Phoenix dactylifera*) are fairly salt tolerant; but none is able to grow with very high amounts of salt. In some landscapes, the salty soils, highly alkaline soils, soils without salt, and leached alkaline soils may be *very* spotty, in areas of 5 to 10 square meters.

With modern soil surveys, supported by good laboratory services such extremes of soils can be avoided. The soil needs to be examined to considerable depth, not only for salt accumulations but also for claypans or other hard layers that interfere with drainage. With good drainage, and irrigation with water of low salt content, the salts can be leached out and not allowed to accumulate (Figure 34). With good leaching the spots of high alkalinity can be improved with additions of calcium sulphate (gypsum) or, with a good supply of calcium carbonate in the soil, by addition of sulphur. The soil organisms change this to sulphuric acid, which neutralizes the sodium carbonate, most of which arises from the hydrolysis of sodium ions from the clay particles in moist periods and is carbonated by the carbon dioxide of the air. As with organic

FIGURE 34

Here is a first cotton crop (*Gossypium hirsutum* L.) on a somewhat salty soil after irrigation and drainage of the Hungry Steepe near Tashkent, USSR.

soils the drainage lines should be closely spaced for best results for reclaiming salty soils. A few examples of using salty soils in several parts of the world are given by H. Greene (1948).

Soil surveys

Soil surveys are vitally needed for agricultural development both for efficient and dependable use in farming and for location of essential transport and other services.

Attempts at extensive development requiring a great deal of money and labor without good soil surveys can be downright irresponsible (Agrawal, 1970). Although I have only high praise for the great British effort in research and planning for the "Gezira Scheme"—irrigation of Vertisols (or Tropical Black Clay) in the Sudan for cotton (Crowther, 1948), the complete failure of the "Groundnut Scheme" in East Africa (Wood, 1950) probably did more to promote soil surveys and climatic studies in Africa than any other event. Unnecessary losses were enormous. So far as I know, no soil scientist visited the area until after the failure and he found that most of the soils were unsuited for peanuts (*Arachis hypogaea*). At first I feared the failure would set back development in Africa. Actually, it gave a helpful scare to many officials responsible for developments.

A soil survey includes those researches necessary (i) to determine the characteristics of soils, (ii) to classify them into defined and named units, (iii) to establish the boundaries of these units and to plot them accurately on photo-mosaics or other accurate base maps, (iv) to predict their response to defined management systems for adapted crops, grasses, and trees, and (v) to predict their stability for highways and other structures.

Since individual soils vary widely in total area, only a detailed soil survey at scales of around 1:15,000 to around 1:31,250, depending on the soil pattern, can be used for good operational planning of farms. Even at these scales some kinds of soil include such small individuals that they need to be combined with other defined soils as "complexes" for mapping units. Such a mapping unit tells the reader that at any exact spot, he may find one of two or three kinds of soil. From the descriptions he can recognize each kind as he walks over the ground within the area shown on the map.

If expensive works for water control are required, their installation may require maps of both soils and relief at even larger scales.

In planning soil surveys one needs to avoid more detail than needed for reliable predictions, otherwise the time and cost are more than necessary. For planning irrigation development and in land reform intended for intensive farming, highly detailed surveys are needed to avoid serious and expensive errors and injustice. Yet it is not prudent to allot funds for such soil surveys over the ones of moderate intensity until the projects are certain to go forward. It is less costly to remap those project areas than to make all the soil mapping conform to the strict requirements for well-planned irrigation on the hazy assumption that they may be needed some day.

The decision on scale of detailed soil surveys, or even reconnaissance soil surveys for broad-scale planning of extensive uses, depends not only on the pattern of kinds of soil and the contrast between them, but also on the investments of funds and labor required per hectare or acre for development.

Yet soil maps of smaller scales are needed for general planning of farming, forestry, and locations of towns and transport. Soil maps at scales of about 1:62,500 or even smaller are useful in planning approximate locations of structures, roads, factories, and so on, for a considerable area. Where detailed soil maps are available the small-scale maps can be made without additional field work. Such maps are essential for the general planning of town-and-country areas.

In seeking suitable new lands for development in poorly explored areas it is commonly considered best to start with a schematic soil map based on available published data and maps with aerial photographs, if practicable. The principle is that each kind of soil is a product of the five factors of soil formation—climate, living matter, earthy material, relief, and age of landform. If reasonable data on these five factors can be had, an estimated soil map can be made with access to a very good library. Work sheets are commonly about 1:500,000 or 1:1,000,000.

With this map, one usually follows one of two courses. Travel routes can be laid out in the area for careful checking of the soils in spots or strips, and along the roads, trails, and water courses. From this field-checked map, prudent selections may be made of promising areas for detailed soil surveys to guide development. In poorly known areas, however, it is more common to choose broad, reasonably promising areas for reconnaissance soil surveys at scales of around 1:250,000. Thus areas designated for detailed soil surveys can be seen more accurately. In heavily forested areas these surveys are made in cleared strips. The late C. F. Charter and his staff developed a method (Charter, 1948) which was used in a recent soil survey in Ghana (Ahn, 1961). Using a similar method of strips, an excellent soil survey for development was made of the southwest region of the Ivory Coast (Carroll and Malmgren, 1967). Leamy and Panton (1966) have also adapted Charter's methods to heavily forested areas in Malaya.

Another excellent soil survey among many for beginning development is one made in western Samoa by A. C. S. Wright (1963). This too may have to be supplemented by more detailed soil surveys as the resources are used more intensively.

Use of aerial photographs

The use of aerial photographs has made it possible to improve soil surveys greatly. They are far less helpful, however, in areas of dense tropical rain forest. Yet some of the most responsive soils for farm development have such cover. In more open areas, streams, roads, houses and the like show up easily and assist the mapper with his location. With overlapping pairs it is also possible to view the relief with a stereoscope and plan routes for mapping. The quality of photographs is increasing continually with better cameras and higher flights.

These photographs should be used where they are available. Some researchers have made fantastic claims for using photographs from planes and even satellites to produce whole soil surveys, without even examining the soils! Such claims are utterly irresponsible; aerial photographs should be used to *assist* in making broad schematic soil maps of small scale. For the most part, the soils themselves do not even show in the photos. Yet the vegetation does and this fact can be helpful for placing boundaries if, and only if, the *local* correlations have been worked out with detailed study on the ground.

The aerial photograph helps the soil scientist a great deal. It helps him keep his location. As he works out the *local* correlations between what shows in the picture and kinds of soil, it helps him place some soil boundaries. In other words, the photograph helps him spread the hard data he has. But with no hard data they do not help him to know the kinds of soils nor where the boundaries among them really come. For example, I have seen clear boundaries in photos between contrasting tropical rain forest; yet one part was the original stand and the other an old second growth, with no related differences in soil. Near the margins of vegetative types small differences in humidity, previous fire, aspect, depth to permafrost or rock, previous use, and so on may change the image in the photograph with or without correlations with soil boundaries. The soil scientist *must* look at the soils to have an accurate soil map to guide operational planning for stable development.

Transfer of data

The basic purpose of soil surveys is to make practical the transfer of data from research and experience safely from one part of a country or of the world to another. To do this, each kind of soil shown on a soil map must be carefully described and named in some standard system of taxonomy. Yet many soil surveys have stopped there.

For guidance in farming the available data from wherever similar kinds of soil are used need to be assembled into interpretations that give responses of the kinds of soil to defined systems of management in quantitative terms of yields of potential crops. Some kinds of soil have many alternatives, others only few.

The hazards of erosion, soil blowing, and slips need to be estimated. Methods of avoiding such hazards and for water control need to be explained along with other critical aspects of management systems. Comparable interpretations for forest and grazing management are needed in many places. These interpretations must give full weight to exotic species that may be considerably more valuable than the native ones.

Potentials for farming vary widely on similar kinds of soil depending on the skills of the operators, the availability and prices of tools and chemicals, and local services. Thus the alternative uses selected by the cultivators may change considerably from the beginning of a development plan as services and market outlets improve, especially on the best soils. The interpretations from the soil survey have use during the process of change.

The interpretations in the soil survey should also include the suitability and hazards of use of soils for buildings and highways. The physical properties of a soil are very important to the stability of highways and other structures built on it (Michigan Department of State Highways, 1970).

For town planning, within town-and-country planning, water control and disposal may be critical. As we warned earlier, if a high proportion of the surface of the watershed of a small stream is covered too thickly with structures and paving, serious floods can be caused because of the greatly increased runoff resulting from reduced percolation of water into the soils.

Remote sensing can be used to get a general appraisal of the soils of a wide area to show where detailed surveys would be most practicable. Then too, we can hope that systems of soil management and potential crops will someday be computerized by the names of kinds of soil. These data could be used for suggested practices wherever like soils exist.

3

Management of soils
for plant production

Perhaps the first requirement of soil management is to provide adequate supplies of food, fiber, and industrial crops for the people of the world at reasonable prices and to provide good labor incomes for the cultivators. The world has an abundant potential for enormous increases in crop production. This potential is not evenly distributed by population. And as my old friend Professor John D. Black warned me many times, people do not take food from a common larder.

In few countries are people comfortable if they are dependent on other countries for nearly all their food supplies. Most countries are best served by some combination of local production and trade for those items they cannot produce economically. A wider range of crops can be produced, of course, in the tropics and subtropics than elsewhere. Yet trade is not easily arranged. Some people in the temperate regions still produce sugar from sugarbeets instead of trading with tropical countries for the more efficiently grown sugar from sugarcane. Yet all sugarcane grown in tropical areas is by no means as efficiently produced as it could be or with as adequate wages for the workers as is desired.

Yet this first requirement for farm products could not be achieved economically without full attention to soil conservation, which means using and managing the soils needed for use in accord with their potentials and in ways that protect or, much more commonly, that increase their productivity and stability on a sustained basis. Some people still think of soil conservation as "saving the soil" as nature produced it! Actually only a few kinds of soil are immediately productive for crops and meadows after simple clearing and plowing; most are not. In practice, soil conservation and water conservation go hand in hand. On most soils, practices are needed to provide the right amount of water through devices for runoff control, irrigation, drainage, or even two or all three of these practices combined.

Most commercial crop plants, including field, horticultural, and industrial crops, are selected for yield and quality, not for their ability to compete with weeds or to grow on infertile soils. Many of our best soils for farming give low yields without lime and phosphate. Nearly all the essential elements taken from soils by crop plants are used somewhere as fertilizer. This includes a few elements essential not to plants but to grazing animals, which get their supplies from plants. In Iceland, for example, the surface material in a large area was lifted and blown by the wind during every windy, dry day. It was called a "humid desert." The whole problem turned out to be an absence of phosphorus. With that element added, grass or small grains were grown and the blowing ceased (Figure 35). Many of the bare slips on unfarmed soils that are not covered with wild plants result from the same deficiency.

FIGURE 35

A moderate application of fertilizer with phosphorus and nitrogen gives good barley (*Hordeum vulgare* L.) growth on an old "humid desert" in Iceland. The bare area to the left has been very large for many years.

FIGURE 36

After fires on old leached tropical soils, trees may be dwarfed from deficiencies of either calcium or phosphorus. Here in this "Cerrado" of Brazil, calcium is extremely low. Courtesy A. C. Orvedal.

FIGURE 37

An experiment with grain sorghum (*Sorghum vulgare* Pers.) on a sandy loam soil in Ghana shows that with nitrogen but without phosphorus (N3-P0) the plants hardly grew at all. The tall ones in the back (N1-P1) did very well where both were supplied.

Yet other volcanic ash may be rich in phosphorus and only nitrogen is needed to get plants growing. Whenever a fresh hill or slope of volcanic ejecta is formed, it is important to make soil tests at once and fertilize and plant seed to prevent erosion sediment from covering land areas below. Apparently in very hot volcanoes the phosphorus and sulphur are volatilized and must be included in the fertilizer, as well as nitrogen, in order to get stable plant cover to prevent erosion.

Thus no skillful cultivator with adequate services is content with the soil the way it is when first plowed. Some farmers have tried and lost both their soil and their labor. Today, nearly all soils being used for crops by skilled cultivators are more productive after farming began than before. But many hectares have been eroded due to poor selection of either the soil or the system of management. Without fertilizers at reasonable prices it may not be possible to have a good system. Taking the world as a whole, phosphorus deficiency is the most widespread and difficult to supply with manure and compost alone; although over a very long time it has been done (Figures 36 and 37). Although deficiencies of nitrogen are widespread, there are several ways to obtain nitrogen. These include some from rainfall near industrial areas and from frequent thunderstorms; from both "free-living" organisms in the soil and symbiotic ones on the roots of many legumes; from manure; and from compost, especially that made of the leaves and stems of high-protein plants.

Data collection

For most kinds of soil, data from research, including outdoor experiments backed up by laboratory study and experiences of users, are available in some country. To make a good start toward agricultural development in many areas, the necessary data must come from more developed areas or countries with comparable soils and growing conditions. Thus for development programs some staff member must be able to find the data that apply to the local soils in the published literature of other countries, a part of which may be in languages other than his own.[1]

For a guide in locating data we need a general soil map of the world at a scale not less detailed than 1:2,500,000. The materials for such a map are available. But it would be expensive to publish. It would need to have a base map showing all the important market towns and villages, roads, streams, and other physical features so the user could locate the experimental stations and areas having soils like those in the area for new development. General soil maps at much smaller scales are available but only a very few of the necessary base data that the reader needs for location are shown. The cost of what is needed would amount to a very tiny fraction of its benefit, if used to avoid the enormous and unnecessary costs of trial and error.

Transfer of skills of plant production

The foregoing may give the impression that transfers of skills are easy. Even at best, with all available data, this is neither easy nor fast because success depends on suitable adjustments to local cultural systems, including education, goals of the people, markets, and services (Kellogg, 1960).

The basic data on alternative systems can be transferred accurately between similar kinds of soil but the development of operating systems must also be adapted to the local skills and culture, hopefully in ways that distribute the benefits widely to both cultivators and to those who find jobs in the essential infrastructure (common services) for agricultural development.

A common mistake in new land development is making the size of holdings large enough for mainly subsistence farming with hand tools. The handicap of small-size holdings has plagued land development schemes for farming in both the developed and the undeveloped countries for many years. Hundreds of papers have been written about it, yet the same errors continue to be made today. The easy way, which nearly always leads to failure, is to establish the size of holdings in terms of land area alone. It is easy

1. Many textbooks and more general "popular" books are translated. But someone familiar with the area's specific needs and some foreign languages must review a proposal for translation to know whether enough actual facts are given to make a translation worthwhile. Mere general descriptions are not helpful. Book publishers still assume, falsely, that all English-speaking professionals read French. Roughly one-half of the most important research results on the tropical soils of Africa are in French and about one-half in English. So a summary book on this subject in French with few or no references in English, or one in English without French references may not be useful. Fortunately the best recent publication on development of tropical soils by Jurion and Henry, summarizing the research and experience from the Republic of Zaire—the old Belgian Congo—has an edition in French (1967) and one in English (1969). See also *West African Soils* by Peter M. Ahn (1970).

for the uninformed to say, "We can give each family 25 hectares," or some other figure. But soils vary greatly in their potential. One needs to consider the productivity index of the soils to arrive at a reasonable size of holding in hectares, and they cannot all be equal. As an extreme example, I saw a land development scheme, with the planning for holdings of equal size drawn in a distant office. Some had responsive soils, many had only hard rock, and most were somewhere in the middle. Of course, it failed.

As the skills of cultivators improve and markets are available for both food and industrial crops they will want to be able to expand the size of their holdings. If this is not foreseen in the original plan and all suitable land is allocated, it may be too expensive or too inconvenient for them to expand. This failure has driven many potential farmers to the big city.

In new developments requiring land clearing, only a part of the basic allotment needs to be cleared at first. Then as skills grow, the holding can be expanded.

Whether starting a development scheme on new land or improving existing peasant farming, the very basis of operation of farms, from primarily traditional social units to economic firms,

FIGURE 38

Elephant grass (*P. purpureum*) is being harvested for dairy cows in Hawaii.

brings many changes, including assistance from an advisory service, credit, and more participation in trade (See Wellhausen et al., 1970, for an example in Mexico).

For success, the principles of economics must be considered from two points of view: (i) The success of the individual farms as firms and, (ii) the wider effects of job creation in the local towns and villages—the Keynesian multiplier effect (Keynes, 1967; Stewart, 1967).

CHOICES AMONG ALTERNATIVES FOR FARMING

As we have emphasized, many kinds of soil can support several alternative farming systems depending on the skills of the cultivators, available inputs and technology, markets, and other services.

Speaking very broadly we find 3 general systems of soil management with widely different yields and labor incomes on similar responsive kinds of soil.

1) The systems developed by indigenous cultivators without the benefits of either modern science or industry. Such systems have been worked out by trial and error over many generations. To one familiar only with modern farming in the developed countries, those systems may appear to be poorer than they really are.

2) Systems which have been improved considerably by modern science but without the associated industrial infrastructure. Better varieties of plants have been introduced that are suited to the local kinds of soil. Composting, with or without animal manure, can be improved by selecting or actually introducing plants with high protein leaves for the purpose. With hand tools, simple water control systems can be demonstrated.

3) Finally we find systems where the farmers have had full access to good transport, electric power, chemicals, machines, and the other products of modern industry. In Figure 38 elephant grass (*Pennisetum purpureum*) has been planted and well fertilized to be cut for intensive dairying in Hawaii. Figure 39 shows excellent elephant grass growing wild and unused, as it does on many good soils of the tropics.

FIGURE 39

Excellent elephant grass is growing wild and unused in a good soil in the western part of the Republic of Zaire. This grass could be used as in Hawaii.

Many transitional systems lie among these three. On many kinds of responsive soils highly productive systems can be used under the third group, with modern infrastructures that would not be useful for farming at all under the first or even the second group. Many of the soils of Western Europe and of the United States that are highly productive for farming would give very low labor incomes under any management system not well supplied with industrial products.

Extensive grazing

Besides farming, or in association with it, grazing and forestry have offered opportunities for rural people for many years. Like farming the alternatives depend on opportunities for markets, improvements in water supplies, and supplemental feed in winter months, dry seasons, or during erratic droughts.

Nomadic grazing without modern facilities for health care of animals and reasonable controls to protect the herbage from overuse can lead to very low returns. In some nomadic societies the prestige of individuals depends on the total number of animals rather than on their condition. Then too, with all the men, women, and children wandering the land with animals, opportunities for education and health care of the people are rarely available.

With some assistance from government for water storage, semipermanent headquarters could be established in carefully selected places near natural water supplies. The water could be impounded for growing emergency fodder and food crops. At such a site at least an orderly market could be arranged where the animals are graded according to quality. Such a market could change the status symbol from numbers of animals to their quality.

On many foot slopes of hills in arid areas, reservoirs may be dug into the deep calcareous strata to hold water above the lower-lying soil to be irrigated. Roman engineers used this method in North Africa. With reserve forage and a market, people could decide to settle down. The women and children could be persuaded to stay there much of the time while the herdsmen are away. Schools could be established.

Such a headquarters could also be helpful for animal inspection for disease. The inability of many herdsmen to recognize animal diseases in time and to get assistance from competent veterinarians have reduced productivity a great deal from what it could be.

Obviously such a plan would need assistance from a skilled staff that understands the local culture and can insure the protection of the people's rights. The staff would have to use unusual tact in discussing settlement with nomadic people.

In all dry areas, the threat of drought and of overstocking are great handicaps of grazing herds. With overgrazing, not only are herbage yields low, but soil blowing can reduce plant cover seriously. With the plant cover reduced, the occasional heavy rain leads to soil erosion. Once this happens, reseeding is necessary along with arrangements for controlled grazing, especially in critical areas; reasonably spaced watering places and possibly drift fencing to avoid heavy concentrations of animals are also necessary. Even in only moderately dry hilly areas, overstocking can cause serious soil erosion that may be costly to correct. For success, control of stocking rates by individuals, by groups, or by government is necessary.

In recent years, machines have been developed for water control on soils used for extensive grazing as well as combinations of chemicals and machines for eradicating brush and other competing plants. In many grazing areas of the United States, for example, the wild fires ignited by electrical storms kept out invading trees and brush. As settlers built roads which also served as fire lines, the threat of fires was reduced. Trees and brush then began to grow on many potential grazing lands. These plants had to be eradicated by some means to avoid their competition with the forage plants. Controlled fires can help eradicate them, yet great care is required to avoid excessive fire.

With irrigation on part of the holding, managers can have both grazing animals and crops. This can lead to a more balanced unit, with needed forage for the dry periods. To achieve success with both grazing land and farm land requires farmers with a wide range of skills.

Forestry

Many farmers supplement their incomes with forestry on soils that may or may not be suitable for farming. In any development project depending partly on such a combined enterprise, a special advisory agent should be available. Not all can be expected to acquire both sets of skills—for farming and for growing and harvesting trees. Ideally, these potentials can be best realized through cooperatives that may purchase suitable tractors and other machines that may be used on a custom basis for land clearing, forestry, road building, and certain heavy jobs on farms. Since many forests grow on hilly or steeply sloping soils, guidance is needed for the location of roads and skid trails.

Many claim horrible results from "clear" cutting—that is from removing all the trees. This is often necessary to replace an old second growth of a poor species in order to replant to a new species that starts poorly in shade. The hazard from clear cutting is increased if followed by hot fires. But on sloping soils a greater hazard is serious soil erosion, with slips followed by gullies, which come about when misplaced trails and roads bring down a concentration of runoff water. An advisor in forestry should understand these soil-water relationships as well as the principles for selective cutting, planting, and harvesting.

APPROACHING THE IDEAL ARABLE SOIL FOR FARMING

Nearly all soils must be changed from the way nature left them to have rewarding commercial farming, especially for water control and for adequate, balanced supplies of plant nutrients. All the nutrients are important, of course, but phosphorus deficiency is widespread. As time goes on more exploration for good deposits will be needed. Many parts of the world have had little searching.

How far to go with such changes depends on needs for specific farm products, their prices, prospects for an adequate agricultural intrastructure, including all aspects of marketing, processing, input manufacture, and especially on willingness to work.

In many places it has been practicable to go a long way by making changes to approach the ideal arable soil—one that furnishes the plant roots a balanced supply of nutrients and both water and oxygen as they are required during the growing seasons. In low-lying fields impracticable to drain, soil material may be brought in from the outside to raise the surface soil above the water table. Others have added soil from the outside to the surface mainly to give a better rooting zone. Some soils can be used effectively only after removal of stones or breaking up of underlying hardpans. A wide assortment of earth terraces (or bunds) are used on many soils to conserve water and guide the removal of excess without erosion. In places one may see great hillsides with level benches held by stone walls and with facilities for irrigation from sources farther up the mountain.

In sandy soils with very high permeability, and where economic returns are favorable, improvements in water supply for crops have been achieved by placing a thin asphalt layer about 50 to 70 cm beneath the surface.

Labor intensive versus capital intensive

One hears these terms a great deal in farming areas with un-employed people. If practices to improve soil productivity require a large amount of labor, usually at low wages, they are said to be *labor intensive.* Yet where tractors and other machines are used with fewer on-site workers to achieve the same ends, the practices are said to be *capital intensive.* Actually it takes a great deal of labor to mine the iron and copper, to make the steel, and to pro-duce the machines. Perhaps nearly the same amount of labor may be used in both. But people forget that most of the "capital" is used to hire labor, perhaps in a distant place, who are fully as much "agricultural" workers as the local laborers. Since these workers are more skilled, their wages are usually higher than those of landless farm workers. Yet in many areas with unemployed people it is better to use local labor, even at low wages, than to im-port expensive machines. Agricultural development is best served when people work efficiently with plans for local manufacturing of as many of the essential machines, chemicals, and other inputs as practicable under competent management. And, of course, as farmers learn to produce efficiently, they need customers—people with jobs who have money to buy. It is most unfortunate that many fail to see the intimate relationships between farming and the industrial sectors of agriculture. Great errors and confusion arise when people speak generally of ' industry" or "industrialization" as somehow apart from agriculture or farming.

In the improvement of primitive farming and in the development of new farming areas, of course, most of the tools and chemicals must come from the outside; but for longtime success the plans for the future should include the necessary education and training to develop as much local production of inputs as is practicable.

Interactions at several levels

The principle of interactions applies in the field. Unless all essential elements for good plant yields are applied together, results can be only fair or poor. We must understand that at the very start it may not be possible to introduce all skills at once. Yet the plan should provide efforts to move in this direction.

As we have already explained, the cultivator probably has unlike soils with different responses to management, all within one holding. These fields ought to be used to give the best combination for the farm as a firm. Some soils that successfully produce forage but not cultivated crops can be used for animals. So the cultivator must think of the balance between forage and feed crops, between industrial crops and food crops, and so on according to his potentials.

Within a trade area what the cultivators can do depends on educational facilities, markets, and other services, including manufacturing and transport. Within a country, an effective dynamic national plan can help improve the opportunities for efficient development of all the trade areas as they interact to give high efficiency for the country as a whole.

Clearing and shaping

Clearing

Until recent years the development of many excellent soils, especially in the humid tropics, has been held back because of the difficulty of clearing the dense forest. Much has been cleared by hand but this is slow, arduous labor. Under shifting cultivation trees are cut in strips, but neither the stems nor the stumps are removed and the rest is burnt within the strips. In the related method of mixed cultures, about one-half of the trees are cut and most of the shrubs are removed.

In these systems all soil preparation, planting, and harvesting is done by hand. But these methods give comparatively low yields per hectare. Certainly they are labor intensive and labor income is low. Jurion and Henry (1969) give interesting details on the gradual introduction of machines to do the job.

In much of the world, stumps were blasted out with explosives after cutting. Then bulldozers were used to uproot the trees, which were pushed into windrows and burnt. In the process much of the surface organic layer was destroyed and even some of the surface soil left in the windrows. Thus the trash did not burn well and these irregular ridges became serious weed patches. Many of the good soils cleared for farming in the Matanuska Valley of Alaska during the 1930's were badly cleared in this way. Such losses of the organic matter are serious in these cold soils because they can be replaced only very slowly.

Now excellent very heavy machines have been developed that sheer the trees off at ground level. This leaves the roots in the soil. A very large, curved, hard steel blade, kept razor-sharp, is pushed by a heavy crawler-type tractor. The few huge trees are first split by a special blade on the main blade and the parts removed just a bit below ground level. (See Figures 40, 41, and 42 for three steps in one system.) In some systems the trees are first pushed over with a heavy steel frame and the heavy wood is crushed. The logs and brush are piled and burned. In areas lacking a dry season, oil is used to start the burning. This process is followed by the use of a heavy cutting disk. Each individual disk is controlled by a heavy spring so that it jumps over the stumps. By this cutting of the surface roots, "sprouting" from the roots is reduced. After the clearing, any necessary terraces or drainage ditches for water control can be made conveniently. With adequate fertilization as needed, the crops can be planted.

This method greatly reduces costs for clearing from roughly $1,200 (US) per hectare to around $160, but not including associated costs for roads and transport. Where roads must be developed, some valuable timber can be harvested and shipped out before clearing (Kellogg and Orvedal, 1969). Obviously such developments must be financed for small holders mainly by the government or by banks or other agencies giving either bilateral or multi-lateral financial and technical assistance.

Shaping and smoothing of soils for use

In the preparation for irrigation, where the soils are reasonably deep over rock, hardpans, claypans, and salty or calcareous layers, the surface can be smoothed or "leveled" for the slow and even distribution of water, either for large areas or between terraces or canals. Depending on funds and labor available, this work can be done with hand labor, draft animals, or machines. Usually the maximum slope, where very heavy rains are not expected, is about 2% for irrigation of cultivated crops or up to 4% for thickly growing crops. The slope should be essentially flat with no pockets or ridges that interrupt the flow of the irrigation water.

FIGURE 40

Cutting a tree in a tropical rain forest at ground level. Courtesy Rome Industries.

FIGURE 41

The "stinger" on the blade splits a large tree in a tropical forest. About three to five splits are made and each is removed separately. Courtesy Rome Industries.

FIGURE 42

Logs and brush are piled after cutting trees in a tropical rain forest. Courtesy Rome Industries.

Similar smoothing is widely practiced with rain-fed paddy rice (Figure 43). This kind of smoothing is done successfully on strongly sloping soils with stone terraces on the contour roughly parallel in strips for cropping as narrow as 1 to 2 meters. Some such strips in Hong Kong, where land is scarce, are only about 0.5 meter wide and held by stone terraces 1 or 2 meters high (Figure 44). In the Mediterranean countries, southern India, and elsewhere on good hilly soils (especially where water is fed by a spring from the upper mountain slope), stone terraces on slopes are a common sight, for fruits and even for vegetables. In many places, where other factors are favorable, the original soil commonly back of the terraces is supplemented by additional soil carried from other spots (Figures 45 and 46).

Some local people who have grown up with this terrace system and a short supply of farm land are skillful at making such terraces, either with stone and mortar, or by simply fitting together large properly broken pieces of hard rock. On some soils, shrubs, coarse grasses, or other plants can be planted thickly on the fairly steep slopes that separate one strip from another and can be depended on to prevent soil erosion.

Yet for all shaping of soils, whether done by hand or machines, it is highly desirable to have the services of a technician skilled in the use of instruments for placing the works and grading the soil at the proper levels for even distribution of the water.

FIGURE 43

Land is smoothed for rice in India.

FIGURE 44

This is an extreme example of narrow, smoothed terraces for irrigated vegetables in Hong Kong.

FIGURE 45

Cultivators add soil to a rice field in India to get enough depth of soil for sugarcane.

FIGURE 46

A cultivator has carted in soil and earth from other places to raise the surface to grow coconut palms (*Cocos nucifera*) already planted at the proper height. In a few years he will have a higher normal level of soil for the palms. Kerala State, India

FIGURE 47

Good papaya (*Carica papaya*) growing on cracked and smoothed recent basaltic lava in Hawaii.

With heavy machines a great deal can be done where other factors are favorable. In an area of high and dependable rainfall, I have seen irregular young lava flows, scarified and smoothed, with holes blasted for fruit trees (Figure 47).

Shaping is also done for runoff and erosion control and reclamation. Some rolling to hilly landforms on which many highly responsive soils may be found are so irregular that once the natural vegetation is removed for farming, the excess water is highly concentrated, resulting in very fast runoff. In these places only small amounts of water soak into the soil for plant roots; in other places water is so concentrated that deep gullies form.

Where the soils are reasonably stone-free and responsive they can be reshaped; the gullies can be filled with soil and smoothed. Then grassed or paved waterways can be provided for the excess water. Soils from loess or old silty alluvium are most easily reshaped for a pattern of use that includes crops, possibly with earthen terraces for runoff control, and semipermanent intensive forage or pasture. With modern earth-moving machines many excellent fields have been made from gullied soils without great cost in relation to the benefits of having good arable soils. If the soils are unresponsive to crop management, simple shaping to control runoff for forestry may be more rewarding.

Similar and more intensive shaping is commonly used for preparing town lots for housing. Success depends on careful study of the soils to avoid impermeable layers that allow water to concentrate and lead to unstable foundations. Where care has not been taken structures have slipped down the slope and been destroyed during heavy rains, especially if aided by even a mild earth tremor. Besides the hazard of unstable foundations, the prevention of concentrated runoff and erosion is essential to avoid damage. Where good sites for building are available, a nearly level well-drained soil is better and safe construction cheaper than on strong slopes or in wet places.

Some soils need shaping for frost protection. Many excellent soils for fruit production are situated on high ground but cannot be used for fruit because of the hazard of frost. Fruit growing here would be feasible only with smooth slopes and U-shaped openings leading down the slope for air drainage. Normally the fruit grower near the margins of belts with frost hazard selects soils that already have good air drainage.

Yet some fruit-growing trade areas, already having the established infrastructure, have considerable areas of suitable soils.

However, such areas are often dotted with cross ridges which prevent the cold air from slipping harmlessly down the slope. If these ridges are composed of deep unconsolidated materials, such as old ocean or lake beaches of glacial drift, it may be highly practical to clear out these obstructions with modern earth-moving machinery. As an example, this technique has been used successfully directly north of Traverse City, Michigan, USA, in an area highly suited to cherry growing except where air drainage is inadequate to avoid frosts.

Many good arable soils have been produced by covering wet areas impossible to drain with good soil, both by hand and with machines, even with coral rock or other wave-exposed rock near the sea. I have seen excellent soils for sugarcane made by plowing the slopes of an adjacent mountain and guiding the resulting sediment from erosion over coral rock with stone baffles.

Other good arable soils have been made by pumping silt out of rivers and placing it back of well-made stone walls (Edelman, 1950). Years ago many productive soils in The Netherlands were made by shipping in sand in barges from dunes near the sea to raise the level of wet soils by 0.5 meter or more.

FIGURE 48

Cultivators have terraced and smoothed the soil for rain-fed rice in Kashmir.

FIGURE 49

Terraced soil for paddy rice is seen here in Assam, India.

WATER CONTROL

Soil and water management, in practice, are nearly inseparable. To use both water and soils efficiently most kinds of soil require a system for runoff control, drainage, or irrigation, or some combination of these. Of course, to benefit from these practices, the soil must have a balanced supply of plant nutrients, varieties of crops that are responsive, and control of weeds and other pests. Only a few kinds of soil are naturally well watered and have adequate supplies of plant nutrients. Investments for good yields are also included in the farming system.

As a first principle for water-control systems on arable soils, the system needs to be carefully adapted to the specific characteristics of the local kinds of soil; otherwise labor and materials may be wasted. Many older abandoned farming communities attest to such failures and the great wastes of human labor.

Runoff control

By far the greater part of the crops grown in the world receive their water directly from rain that falls on the soil naturally. The orderly management and conservation of this rainwater for beneficial use is the major water-control problem for the arable soils of the world.

The productivity of those soils that are reasonably permeable to water and that have variable supplies of rainfall can be greatly

FIGURE 50

Terracing here was done on a very pervious clayey tropical soil (Tropept) in sout
central India. The bottoms of the outlet ditches are on rock. The soil is so perv
that water moves down into the lower soil and seeps away without erosion.

FIGURE 51

A stonewall terrace on granular Oxisol, made without mortar from carefully cut
and fitted stones, roughly on the contour, is seen here in Kerala State, India.

FIGURE 52

Low terraces west of Volgograd, USSR, are used for a wide band of shelter-belts to protect the city from the dust of soil blowing during dry periods.

improved by several kinds of terraces. These are laid out in such a way, usually at a slight angle to the contour, that the runoff water is arrested in its movement down the slope and thus given more time to penetrate the soil. Enormous areas of soil in the world, especially in southern and southeastern Asia, benefit from this practice if terraces are well laid out and provided with proper outlets for the excessive water (Figures 48, 49, 50, 51, and 52). If they are poorly laid out, poor drainage from some combination of waterlogging and soil erosion are likely to result in wastes of resources.

Runoff control is basic to soil erosion control. Earthen terraces themselves need to be protected by thick growing vegetation or in other ways where rains are heavy. The aim of the system is to prevent concentration of water, to encourage its penetration into the soil, and to provide that any excess moves off slowly into prepared outlets.

Many potentially excellent soils fall into this category, especially those that are reasonably permeable to water, have long, gentle slopes, and where rains come unevenly. When soils are well drained the total amount of moisture for plant roots is increased. The roots have not only enough water during rainy seasons but also enough water is stored in the rooting zone for a crop during the drier season as well. In many situations, the ground water is recharged so that water is more abundant in wells in the lower slopes during the drier seasons when rains are few.

I have seen this practice alone double yields in India and elsewhere. Then as the systems were supplemented by fertilizers, better varieties, and pest controls, production exceeded that in traditional systems several times.

Opportunities for success with properly designed systems are widespread, including great areas in the less developed countries that have abundant labor and little capital. Except for the design and layout by a skilled engineer, all the work can be done by local people, in great contrast to large irrigation schemes. The earth can be moved in several ways. Some of the best operating earth terraces I have seen were built with shovels and the earth carried in baskets on the head. Some would call this "labor intensive," since all the wages are paid locally. Bullocks or other draft animals can be used with simple graders. Where labor is scarce tractors and machine graders are generally used. Unfortunately, some countries have used this expensive equipment where handwork could have been used; and yet the unemployed people just watched and the money left the area!

In many village areas where a runoff-control system would be very useful, the land has been subdivided into many tiny parcels. Traditionally the people have had terraces or bunds on their property lines. Nearly all attempts to put terraces on old property lines lead to both waterlogging and erosion. In fact, for efficient water control it is necessary first to develop a scheme for consolidation of the holdings, according to groups of similar soils. This can be done in such a way that terraces, irrigation canals, or drainage ditches can be laid out properly. At least a part of the terraces may need to be on new property lines. (See page 131.)

Irrigation

Irrigation is one of the oldest water-control practices. In parts of the Mediterranean region and especially east of it, irrigation became a necessity when the area became drier following the end of the Pleistocene. But as the early history of the Middle East shows, however, military control is necessary. With the destruction of dams and canals, sedimentation and salt accumulation make serious trouble.

Most large irrigation schemes require large dams on upland streams with the water distributed through systems of canals to the farming areas. Yet much soil in the world is irrigated from wells and distributed by small canals. Some still use the old Persian wheel or hand lifts (Figures 53 and 54). With the increasing development of hydroelectric power, electric motors are used. Others use motors operated by gasoline or diesel fuel.

With an effective system of irrigation and a good source of water, responsive soils of deserts and semideserts can be farmed efficiently. The productivity of other soils in normally humid and subhumid regions can be made more productive by supplemental irrigation during either regular or irregular periods of low rainfall. Such irrigation, especially with sprinklers, has been increasing in several parts of the United States, Europe, and elsewhere on permeable soils. Even in such areas good design is critical to avoid erosion and waterlogging. Should a heavy rain fall directly after irrigation, and the excess water cannot soak through the soil and if there is no effective drainage system, crops may do poorly.

We have already mentioned the advantages of shaping the soil to make it smooth and of leveling the surface for border irrigation. Since this cannot be done well on soils that are thin over rock or that have lower layers of hardpans, highly calcareous or salty materials, laterite, or gravel, soil selection based on a highly detailed soil survey is essential to avoid costly errors.

Perhaps the importance of leveling for irrigation needs emphasis. Soil that appears to be nearly level to most people actually has many low swales and mounds. When this is so, the new irrigator is likely to put on far too much water. In trying to wet the "high spots" he may waterlog the low ones (Figure 55).

All expensive developments should be based on detailed soil surveys. Certainly the risks of not doing so are especially great in irrigation systems based on large storage dams and canals. With good soil surveys the potential yields for each kind of soil can be estimated closely. With the detailed soil map and a detailed topographic survey, the roads and canals can be located in the best places. I have seen many failures due to bad location of roads, railways, and irrigation canals built without deep culverts; poor placement prevented drainage and resulted in hopeless waterlogging of fields. Then a classification of the land for irrigation possibilities

FIGURE 53

The cultivator in western India made this "Persian" wheel for irrigation, with wooden gears, clay pots, and all.

FIGURE 54

Here is an example of lift irrigation. The worker thrusts the bucket into the well and lifts it out with the aid of the weight fastened to the far side of the cross pole. Kerala State, India.

considering yield potentials, management responses, size of holdings, and practicability of water delivery can be made.

Very costly systems that depend on large dams and canals to supply irrigation water are seldom economically feasible unless the system is very well designed and installed. This includes the layout within fields and surface leveling, along with terraces, drainage, and salinity control where needed. That is, irrigation should aim to remove deficiencies and excesses of water. Then too, large investments in fertilizers, the most responsive kinds and varieties of crops, and pest control are necessary for the high production required to pay the costs.

I have seen soils irrigated at great cost, but without shaping the soil properly and without drainage so that yields were actually lower than on adjacent unirrigated land. The irrigated soils must yield a great deal more to pay for the investments and extra labor.

In some areas the management of the large dam used for water storage should give emphasis to hydroelectric power for farm use, including the pumping of good quality ground water from tube wells. In all irrigation, close attention is needed in advance to the quality of the water and the hazard of salinity.

FIGURE 55

A poor crop of berseem (*Trifolium alexandrinum*) has been planted in ill-drained, irrigated land near Karnal, India.

Another handicap to large irrigation schemes is that usually much of the cost must be paid from funds outside of the local area. The advantages of yield and income need to be substantial for such allocations to be received.

Many small irrigation developments have been made by individuals based on small impoundments behind dams or in hollowed out storage basins a little way up a mountain slope. As mentioned under the brief section of grazing, early Roman engineers did this in North Africa. Many soils of arid and semiarid regions have accumulations of hardened calcium carbonate beneath, commonly called *hard caliche* or *croûte calcaire*. If the material dug out can be taken up through a fairly narrow slit, evaporation is much lower than from an open pond.

Drip irrigation

Drip irrigation has been coming forward as a precise method of adding water where it is scarce. Pipes are placed on or in the surface soil with holes by the individual plants. The Israelis have developed useful outdoor schemes based on this system.

Carl N. Hodges and his staff, of the University of Arizona, have developed successful schemes for drip irrigation of sandy desert soils under plastic. Waste heat from engine-driven electric generators can be used to desalt seawater. In cooperation with the University of Sonora, in Hermosillo, Mexico, a pilot scheme was established at Puerto Peñasco, Mexico. Besides furnishing fresh water for the town, a part was used to grow vegetables under closed, plastic greenhouses. The inside air is kept humid. In cool evenings some water condenses and can be reused. In addition, carbon dioxide gas, obtained by scrubbing the exhaust gases from the diesel engines, can be added to the inside atmosphere for improved growth. Other essential plant nutrients, hormones, and so on can be added in the drip irrigation water. Another comparable facility is underway in Abu Dhabi (Hodges, 1969).

For areas lacking fresh water but with smooth sandy soils and other resources, including petroleum at relatively low cost, this scheme may offer relatively cheap sources of vegetables, fruits, and perhaps other crops.

Drainage

Systems of artificial drainage have been installed in many of the most productive arable soils in both humid and dry regions of the world. In much of the United States and Western Europe such systems, with tile lines beneath the surface, are not even noticed by most travellers unless they see large open drains or ditches that receive the water and carry it to a river, lake, or the ocean. Now many of the main drains are large covered tile because of the greater convenience of field operation as compared to open ditches.

The design may be simply to supplement the natural drainage ways and to tap seepy areas. For large, nearly level soil areas, lines of tile are laid out in a grid pattern. Some systems are a combination of both of these. As in other water-control systems, engineers skilled in the use of instruments are necessary for the design to be certain to have the tile or open ditches at even, gentle grades and properly spaced.

The permeability of the soil to water has a great influence on design of drainage systems. In soils of low permeability water moves through the soil slowly and the drains must be more closely spaced than in soils of high permeability. The usual heavy rains and possible flooding as snow melts must be taken into account in planning the system to remove excess water, including the large ditches and outlets.

Between the drains, open or closed, the *water table*—the upper limit of the part of the underlying soil that is saturated with water— is an arc, not a straight line. Thus to be effective, the highest part of the arc between drains must be below the normal rooting zone of the plants. Also the principal crops to be grown must be taken into account. Commonly it is not practical to grow perennials, such as fruit trees, on soils that can be drained well for annual crops.

Many naturally wet, highly organic soils can be made into good arable soils. Others are too acid, a few are too alkaline beneath, and many are too cold. In cool-temperate climates, crops on mineral soils are less subject to frost than those on organic soils, even side by side. Besides the principles of drainage design already mentioned, those for organic soils include the estimation of the amount of settling of the soil after drainage, prepared for cropping, and as use goes forward. If the depth to underlaying earthy material is very irregular, drains may settle unevenly for a few years.

Some drainage systems for responsive organic soils have not functioned well because the drains were placed much too far apart. The middle portions between drains were still too wet, while near the deep drains the organic matter oxidized more rapidly and settled more. As a result the fields became uneven.

Even in middle latitudes, the organic matter of drained organic soils oxidizes faster than it is commonly practicable to replace it. All drained and cultivated organic soils in warm temperate and tropical climates will ultimately disappear. Yet this does not mean that it is impracticable to drain and use them for valuable crops (Skoropanov, 1961).

Salinity control

In the previous chapter the nature of salty soils and the basic principles of their reclamation were explained. This conclusion of the discussion on water control is intended to remind readers of the importance of avoiding the use of salty water in irrigation. In laying out designs for any water-control systems, the hazard of including salty soils, perhaps inadvertently, must be kept in mind. Many of them can be used with high quality water, adequate natural or artificial drainage, and appropriate chemical treatments.

Many irrigated soils in semidesert areas also need drainage to remove excess salts. Here too some have put the drainage ditches or tile lines too far apart, such that the salts between the drains are merely "pushed" down a bit into the soil but not moved out.

Dry farming

Between the soils that have adequate moisture and runoff control for some crops at least, and the dry soils that must be irrigated for any crop growth, are soils of intermediate moisture used with what have been called "dry-farming" systems.

At one time a simple system consisted of 1 year of a drought-resistant crop, such as selected barley, sorghum, wheat, or perhaps maize, and a following year of summer fallow, with enough cultivation to kill weeds. Moisture was conserved during that season. Some aimed to have a rough surface with clods or with ridges made with a lister to help control soil blowing.

Although this system is still used, most cultivators now take measures to control runoff with graded terraces, minimum tillage during the fallow year, and other measures to protect the surface against the direct force of high winds when the soil is loose and dry. Some also leave strips of nonweed tall plant stalks to catch snow, while some leave alternate strips unplowed.

Now most farmers on soils of intermediate moisture avoid excess tillage, emphasize surface water control, and select varieties

of adapted crops for drought resistance. The principles are similar
to those followed elsewhere but put together to emphasize water
conservation through weed control, with or without chemicals,
runoff (and snow) control, minimum tillage, and protection against
soil blowing.

TILLAGE

The hand hoe of stone or iron with a wooden handle and a simple
stick on a beam pulled by man, woman, or beast are among the
oldest tools of farming (Figure 56). In the early eighteenth century, on the basis of a false theory, Jethro Tull in England invented
the *horse-hoe*—a horse-drawn cultivator. He held that it was necessary to keep the soil pulverized so that the fine particles of earth
could enter the "lacteals" of the roots. With grain planted in rows
and cultivated, he had large increases in yields without additional
manure. We now know that the reason was not that soil particles
entered the roots, as he supposed, but because he killed the weeds,
which had become very bad in English grain fields. In Japan and

FIGURE 56

An old type of locally made plow pulled by bullocks is preparing land for
paddy rice near Karnal, India.

elsewhere cereal grains are now planted in rows for weed control by cultivation (Figure 57).

Yet as a result of Tull's experiences, many farmers till the soil too much. I did when I farmed as a boy in the early part of this century. We have learned that excessive cultivation, especially when the soil is moist, tends to destroy the natural crumbs and to make the soil massive. Such soil breaks out in clods that roots penetrate with difficulty.

The ideal arable soil has a high proportion of the mass in crumbs. We say that such a soil has good tilth. Water penetrates and roots extend themselves easily. But we do need to control weeds; here chemicals have helped recently. Once weeds are reduced and kept from seeding, control is not too difficult.

So now we say—plow and till well at the proper time, and as little as necessary; kill weeds, develop good seed beds, and make the soil receptive of the water that comes as rain or by irrigation. Then too, in soils subject to summer drought, it is well to place fertilizers deep, especially the phosphatic fertilizers, so that there may be enough moisture for the roots to take in the nutrients where there may be too little in the upper 20 to 25 cm.

The moldboard plow is a useful tool to plow under applied manure, compost, and other organic residues. But the organic

FIGURE 57

Barley is planted in rows for weed control in Japan. Pears (*Pyrus* L.) in the background are trained on large arbor for easy hand care.

matter should not be all completely buried. The pressure of plowing in some soils tends to compact the lower soil into "plow pans," especially with repeated deep plowing when the soil is moist.

Then too, we need to avoid leaving the soil bare as much as possible. Exposure to beating rains and to hot sun tends to destroy the crumb structure. On sloping fields in humid regions with sharp showers soil erosion is promoted and in dry areas soil blowing is increased.

Organic matter, and the microorganisms and small invertebrates such as earthworms that live on it, promote good structure and soil permeability. Lush crops tend to protect the surface soil from rain and sun and leave more roots in the soil. A poor maize crop promotes erosion; a heavy crop protects the soil from beating rain drops and has stronger roots to reduce runoff.

In dry areas overtillage tends to lower organic matter and lay the soil open to soil blowing. Efforts should be made to leave stubble in the surface rather than to bury it. To get the most rain to enter the soil, it is helpful to carry on tillage on the contour where practicable.

The weight of heavy machinery tends to pack some kinds of soil, especially when they are moist. The load of the machines may be transferred to softer layers beneath and pack them. The movement of water downward or laterally to drainage tile is greatly reduced. Oxygen, essential to roots, is crowded out. In northwestern parts of Europe, this loss of open soil structure with abundant soil pores leads to difficult problems of tillage and drainage for good yields. This internal packing can be partly or wholly avoided by rotating deeply rooted legumes, such as alfalfa (*Medicago sativa* L.), with the cultivated crops. Yet some farmers cannot make use of the greater amount of forage without major economic adjustments. Formerly most farms did have livestock and lighter machines; now more specialize in one or only a few crops, without livestock.

Many kinds of soil that are not especially wet do have claypans naturally, or they have developed or become worse as a result of tillage. For these soils a tractor-drawn deep chisel to a depth of 30 to 50 cm can be used to break them up. At the same time deep fertilization can promote strong root systems that aid in preventing compaction.

Recently in the United States and elsewhere, much emphasis is being placed on minimum tillage or mulch tillage to promote soil structure and water movement into the soil and on avoiding soil erosion or soil blowing after a crop has been harvested. Many sorts of machines are used. Some till the soil only in a narrow band where the seeds are to be planted. As required, chisels can be added to break up the lower soil. The straw and other residues from harvesting a good crop of wheat or maize, left on and in the surface soil, can amount to 3,000 to 6,000 kilos per hectare. For good results, one may need to apply a bit more nitrogen fertilizer to balance the carbohydrate of straw. The residue can delay the warming of the soil in the spring of mid-latitudes. Also it may require special measures for weed and pest control. It does reduce the total work of tillage (Hayes, 1971).

The tillage systems of cultivators in the world vary widely and will continue to do so for a long time, depending on the local kinds of soil, the rotation of crops, the hazards of weeds and other pests, the relative availability of labor and machines, hazards of soil erosion and soil blowing, and so on. With the many new potentials, advisory services could well give this topic a high priority for local adaptive research.

FIGURE 58

Animal bones are going to market in India. For a long time bones, and their phosphorus, were shipped out of the country.

FERTILIZERS AND SOIL AMENDMENTS

Nearly all soils require additions of plant nutrients to become productive for good yields of high quality crops, and especially to remain so after nutrients have been removed by crops and livestock, although there is little direct relationship between additions, removals, and yields except over long periods. Some farm areas near the port cities were depleted in phosphorus by export sale of bones over many years (Figure 58).

Some kinds of soil have abundant supplies of phosphorus, calcium, or other nutrients. Then too, much depends on the return of crop residues, compost, and manure. We have already mentioned that the old *balance-sheet* theory—we must put in what we take out—is a poor guide to current needs of nutrients.

The elements taken from the soil as plant nutrients are somewhat arbitrarily listed as "primary" and "secondary." In the first group are nitrogen, phosphorus, potassium, calcium, magnesium, and sulphur. The second group, commonly called micronutrients, include iron, copper, zinc, manganese, boron, molybdenum, cobalt, and chlorine.

Except possibly for chlorine, important additions of each of these are made by commercial farmers on certain kinds of soils for some crops. Since magnesium is essential for the movement of phosphorus within the plant, it is especially needed for oil crops, such as African oil palm (*Elaeis guineesis*). Harler (1971) has pointed out the need for sulphur and some of the trace elements in tea gardens in several parts of the world. He attributes this growing problem in tea (*Camellia sinensis*) as due in part to the decreasing use of synthetic fertilizers made from the "impure natural salts," which had the secondary elements as "impurities," such as zinc, copper, manganese, boron, and molybdenum. All of these secondary elements are important in fertilizers for some soil-crop combination in the world. Then too, most soils adapted to tea culture have been well leached and are acid. I have seen areas used for legumes nearly barren simply from the lack of a few kilos per hectare of boron or of molybdenum, for example. Many soils cannot support either sheep or cattle without small additions of cobalt. Calcium is added in limestone and both calcium and magnesium are added in dolomitic limestone. As pointed out earlier, however, some of these nutrients are added naturally from dust and in rain.[2]

Although it furnishes calcium or calcium and magnesium, limestone is used on soils mostly to correct acidity. For most crops and microorganisms in temperate areas the ideal pH approximates 6.5. It can be down to 6.0 or ever lower in tropical soils. Some crops, such as tea, blueberries (*Vaccinium* sp.), and cranberries (*Vaccinium oxycoccus* L.) require a lower pH. In many old acid soils, an application of lime to raise the pH suddenly to 6.5 may cause problems of deficiencies of the micronutrients, especially in fruits. If so, calcium as a nutrient can be added as gypsum and magnesium as epsom salt. Yet generally speaking acid soils need to be limed for most farm crops before good results can be had from the other fertilizers. As we mentioned earlier, many soils in the rainy tropics need additions of magnesium for the phosphorus to be moved into the seeds of plants, such as African oil palm.

In the early 1930's several of us calculated (in an unpublished report) that on acid soils in the United States, farmers would have better yields by following the advisory service recommendations on lime, with no increase in their fertilizer, than they would by following the recommendations on fertilizer without increasing their lime applications. Usually lime does not give obvious results so quickly as nitrogen fertilizers. Fortunately, most farmers in the United States have since learned this.

Most of the microorgansims that fix atmospheric nitrogen and those that decompose organic matter thrive best at around pH 6.0 to 6.5. Phosphorus and other nutrient elements are most nearly available at that pH. At low levels of acidity, aluminum be-

2. This is hardly the place to go into all the details of the functions and interactions, with one another for various kinds of plants, of each of these nutrient elements essential to soil productivity. (See Ignatieff and Page, 1958; U.S. Department of Agriculture, 1957; Buckman and Brady, 1969; Russell and Russell, 1961 (new edition in press); and Donahue et al., 1971). Recently the Soil Science Society of America has published a large book *Fertilizer Technology and Use*, which summarizes recent research and experience. (Olson, 1971). Also the Society has published another large and highly technical book by several authors on *Micronutrients in Agriculture*, including their relations to soil fertilization, animal health, uses, and hazards. (Mortvedt, 1972).

comes more soluble and is toxic to some crop plants. At high pH levels, alkaline soils have poor nutrition for most crop plants.

Decomposing organic matter naturally produces some acid; so does the sulphur added by rain passing through heavy sulphur fumes. Soils in humid areas lacking any such rain become deficient in sulphur as a plant nutrient. In such places it turned out that the sulphur in superphosphate was giving the increases in yield, not the phosphorus. Some of the nitrogen fertilizers, such as ammonium sulphate and a few others, tend to produce some acidity. Many of the most productive soils of humid regions require occasional additions of finely ground limestone.

Besides the nutrient elements listed earlier, plants also take in other elements if reasonably abundant, including silicon, aluminum, selenium, and even the heavy metals.

At one time, certain volcanoes gave out a great deal of selenium that settled on the surface of rocks and soils. In humid regions much of this moved on out. Some became concentrated in muds that later became shales. Plants growing on soils developed from these shales in areas of fairly low rainfall contain much selenium. Forage plants grown on such soils are toxic to grazing animals and food crops are toxic to people if these make up a high percentage of their diet (Byers et al., 1938). They also reported that Marco Polo's observations during his journey through Asia include the typical symptoms of selenium poisoning.

Later it was discovered that at least some selenium is essential to sheep and cattle. Work has shown that the white muscle disease of lambs and calves is rarely reported if the selenium content of the forage is about 0.1 part per million (Kubota et al., 1967). The Soil Bureau of New Zealand has done much work in this area of the influences of elements on animal nutrition and has found that the grazing animals also take in selenium and other elements directly from the soil adhering to the vegetation (Healy et al., 1970). For a detailed treatment of the influence of the trace elements on human and animal health see Underwood (1971).

Future research in this area will also deal with lead, chromium, vanadium, cadmium, and other elements. Iodine deficiency in foods and feeds is harmful to both animals and humans. With the increasing accuracy of analytical methods for small amounts, with improved sampling methods, and with more refined clinical observations, the kinds of soil, the kinds of crops, and soil management practices will be more precisely defined for the benefit of both human and animal health.

Manure, compost, and wood ashes

These were the traditional sources of plant nutrients for early farming and are still widely used by some.

Manure

Since early times animal manure was conserved and spread on the land both for its fertilizing value, and for its promotion of good soil structure when incorporated into the soil. It contains nitrogen, phosphorus, and small amounts of the other essential plant nutrients. Homer sang how Odysseus, on returning home from his wanderings after Troy, was recognized only by his old dog lying on a heap of dung, "with which the thralls were wont to manure the land."

Most cultivators still use it, some with great care to save both solid and liquid parts without excessive leaching (Figure 59). Yet many in south Asia use much of the cattle manure for fuel (Figure

FIGURE 59

Pictured here is a common barnyard scene in central Europe (Yugoslavia) with well-preserved animal manure and pit for the liquid to be pumped out.

60). Recently as fertilizers became cheaper in areas of high labor costs, some farmers dispose of manure in other ways than application to soil for its plant nutrients and organic matter.

Compost

Compost is used in areas where little manure or chemical fertilizer is available. Wild or cultivated plants that grow well and have leaves with a good green color, obviously have few nutrient deficiencies. Such plants, especially those with high protein leaves, are excellent for composting. They are likely to have supplies of both nitrogen and phosphorus, the two most commonly limiting elements for food crop production. The small branches of some trees and shrubs are especially good. If animal manure is available, it should be added to the compost mix (Figures 61 and 62).

The purpose of composting is to reduce the ratio of carbon to nitrogen. In a productive soil this ratio is about 12, that is about 12 carbon atoms for each nitrogen atom. In green plants the ratio is commonly considerably higher and in dry leaves and straw much higher.

If such material is mixed directly in the soil, considerable moisture is required for decomposition. The population of micro-

FIGURE 60

Cattle manure in many countries of southern Asia is prepared in cakes, dried, and used as household fuel for cooking. They make an even, slow fire.

FIGURE 61

Compost is made in a group of pits from waste banana leaves and cuttings from a legume shrub (*Flemingia congensis*) in western Zaire. The pile had just been taken out and the remaining material will all be turned through the three other bins. This is excellent material to bury near individual banana plants.

FIGURE 62

Compost is made from sugarcane refuse after milling in India.

organisms increases enormously with all the fresh carbohydrate for food; they compete directly with the crop plants for phosphorus and nitrogen. This can be shown easily by putting some sugar on a soil in grass or barley; it causes a depression in yield. But later the same spot may be more productive. The sugar supply is then exhausted and many of the microorganisms die, furnishing more nitrogen and phosphorus to the plants. This effect is helped by the fact that some of the stimulated microorganisms are able to take part of their nitrogen from the air and fix it in compounds that plants can use.

Many cultivators use compost but some others need guidance. If their low yields are clearly due to phosphorus deficiency, there is little use of recommending superphosphate if they live say 400 kilometers from a port city and perhaps 100 kilometers from a highway. They can be shown how to select high protein leaves from the wild trees and other plants, or even to grow such plants, for composting. In the compost they can mix in any animal manure and kitchen refuse. Plants with high protein leaves are likely to be able to take phosphorus from the soil.

I have seen excellent crops of bananas grown at M'Vuazi in the lower reaches of the Zaire (Congo) River Basin using compost made from manure, banana leaves, and *Flemingia congensis*—a deeply rooted woody legume—with a surface mulch of cut elephant grass (Kellogg and Davol, 1949 and Figure 61).

Some have said that for the world as a whole, nitrogen is the most common nutrient deficiency. I should put phosphorus deficiency as more serious, however, because it is easier for poor cultivators to find the local materials to correct nitrogen deficiency than to supply phosphorus. In places with many electrical storms considerable nitrogen comes in the rain. It is fixed by organisms living symbiotically on the roots of clover (*Trifolium*) and other legumes, although not significantly on some woody legumes. Then some free-living organisms in a good soil also fix nitrogen.

Although the sources of nitrogen are explained in great detail in most standard textbooks, very much remains to be learned. I have seen many instances of good maize and other crops, free of nitrogen deficiency, without knowing why. Then to add to the puzzle I've seen fair crops of maize grown by nearly isolated cultivators on sloping soils from basalt, while nearby on soils from granite they had poor crops, with the maize plants showing severe nitrogen deficiency. In neither place did they use fertilizer, animal manure, or much compost.

In fact a major research need is to know more about the sources of nitrogen in soils, especially in tropical areas. I suspect that the principle unknown sources are unidentified tiny animals, fungi, or bacteria. If we knew the exact relations, management systems could be worked out to increase the amounts. We know now that the fixation of nitrogen by algae has kept many paddy rice cultivators alive for a long time. And they also fix nitrogen in desert areas.

Animal manure has both nitrogen and phosphorus. It should be added that a few kinds of soil in the tropics, growing plants with phosphorus deficiency do not respond to ordinary phosphatic fertilizers in the absence of manure. Research is urgently needed on the causes. As only a wild guess perhaps, the manure serves somehow as a chelating agent.

Wood ashes

Wood ashes serve to furnish especially potassium and calcium along with small amounts of other mineral nutrients. Many use them for fertilizing coconut palms, perhaps mainly for the potassium. Wood ashes also serve to help correct soil acidity. As we have seen their production through fire is a critical aspect of shifting cultivation in forested areas.

Chemical fertilizers

Nearly all fertilizers and soil amendments, such as lime and gypsum, have an effect on soil productivity beyond the current crop. The direct effects of the supply not used in the year applied are called *residual effects.* Crops a second or third year later may use additional phosphorus left over in the soil from the crop to which it was applied. A fertilizer used on a crop may result in much larger root systems and better tilth, and thus in more food for nitrogen-fixing organisms.

Curiously, many in the fertilizer trade use the misleading term "fertilizer consumption," in reporting the amounts used. Actually, considering the direct and indirect effects that follow application, a more nearly appropriate term would be *investments in fertilizers* or simply *fertilizer used.* Also many use the misleading term "plant food" for fertilizer. The materials that nourish the plant in the sense of food are organic. The plant takes in *nutrients* from the soil, air, and water and, with the energy from the sun, makes its food. The food stored in the fruits, leaves, roots, seeds, tubers, and stems becomes food for the plant, animals, microorganisms, and us.

Use of soil nutrient tests

The use of soil nutrient tests as a guide to fertilizer needs is recommended as one means for suggesting the kinds and amounts of fertilizer to use. Foliar symptoms on growing plants are also very useful, perhaps especially for the micronutrients. New research is going on in this field in several countries, but one of the best guide books is the one by the late Professor T. Wallace (1951).

The interpretation of chemical test results vary widely with contrasting kinds of soil. A test result for potassium, for example, may indicate that one kind of soil has ample supplies for most crops but a contrasting kind of soil showing the same test may be too low for satisfactory yields. Even a few soil scientists still fail by giving the same recommendations based on similar values of soil tests for contrasting kinds of soil. Then too, the previous yields and anticipated cropping systems are helpful in making interpretations from test results.

Appropriate soil tests can be very useful. Knowing the kind of soil, the yield and appearance of the previous crop, and the previous management, and knowing the crop to be grown, a soil scientist could make a more nearly valid recommendation than he could from the soil-test results alone. *But no such choice needs to be made.* With the soil-test results *and* the other information he can give the most reliable suggestions.

Field experiments

Field experiments can be basic guides for fertilizer responses in controlled systems of tillage, cropping, and fertilization in combination with other practices. Yields can be measured accurately. Such results can be used to standardize field tests, if the field experiments are established on representative areas of known kinds of soil. Each useful field experiment is a fair sample of one kind of soil extensive enough to be worth sampling. Yet it is not uncommon to see this principle neglected with boundaries between contrasting kinds of soil running through the plots or to see plots on areas of soil that are rare. Of course, results from such field experiments are not helpful. Field experiments are costly to establish and to operate accurately. But much has been learned from the good ones that guides farm development and improvement on the kinds of soil represented.

Fertilizer needs in little-known areas

Field appraisal of fertilizer needs in a little-known area can be carried on quickly for an approximation of fertilizer needs. A carryall truck can be fitted out with small bags of the principal fertilizers for a standard plot. By working with the village headmen, small plots can usually be arranged for in typical fields of cultivators. Estimates can be made by a trained man of what to try. These could be 1-N, 1-P, and 1-K, 2-N, 1-P, and 1-K, 1-N, 2-P, and 1-K, etc. If observations suggest possible deficiencies of calcium or magnesium, they may be added as 1-N, 1-P, 1-K, and 1-Mg, or other ratios.

With standard containers and equipment, numbers of such plots can be laid out rapidly. Instead of measuring yields a technician can visit the plots frequently to measure height and examine the foliage. If leaves are turning yellow or are otherwise off color, different strips of the growing crop can be sprayed with dilute solutions of zinc, ferrous iron sulphate, manganese, copper, or whatever seems most likely to be deficient among the micronutrients.

At harvest time the yields of the plots and subplots can be estimated. Without doubt some will show substantial differences over the rest of the field. On a well-managed modern farm, one could hardly estimate closely enough to distinguish clearly between yields of wheat, say of 30 and 40 quintals (1 quintal = 100 kilos) per hectare; the cultivator, though could easily see the difference between 7 and 14 quintals per hectare.

The results from such plots give the advisor a working concept of the probable nutrient deficiencies and responses of fertilizers. The same plots also serve as demonstrations. Unless the peasant cultivator can see a large difference he is unlikely to be impressed. Such results over 2 or 3 years, while by no means final, can serve to suggest trials in more carefully attended experimental plots.

Fertilizer applications

Fertilizers can be applied in several ways—broadcast by hand, drilled in with a machine, or placed just below and at the side of the seed or tuber. These methods are explained in many booklets and standard texts. Local application of phosphatic fertilizer is especially useful to those soils that tend to fix it in forms not available to plants. Such soils are common in warm temperate and tropical regions. Individual applications are best for tree and shrub crops. In this way in a large planting, allowance can also be made for differences in soils.

Broadcasting fertilizer on the surface on sloping soils may be wasteful if exposed to hard rains that wash it off. This is especially likely in mid-latitudes if fertilizer is spread at the time the soil freezes in autumn.

Effect of one crop on a subsequent crop

The effect of a crop on the one that follows can be to improve or to reduce soil fertility, aside from the considerable influence of fertilizers applied earlier. Alfalfa (lucerne) and sweet-clover (*Melilotus indica* L.), for example, have strong roots and subsequent grain crops are commonly better. The organisms on their roots fix atmospheric nitrogen. Thickly-growing meadow crops tend to suppress weeds. Many of the plants that have become weeds were, in natural conditions, the invaders that could get started only on bare soil. The crop residues add organic matter. If these are rich in carbohydrate and low in protein, the effect may reduce the nitrogen available temporarily. But additional nitrogen fertilizer prevents this and the tilth of the soil is improved.

Continuous growth of shallow rooted plants favors the formation of claypans and plowpans in some kinds of soils. Yet if strongly rooted legumes are grown in the rotation of crops, these effects may be largely avoided.

SOIL EROSION CONTROL

As we saw in earlier chapters, soil erosion is a natural process of the movement of earthy material down slope by running water.

Actually a high proportion of the food crops of the world are developed from water-laid sediments from erosion in the past, along streams and in their deltas. In many hilly landscapes, erosion continues actively. Accelerated soil erosion does take place on sloping soils used for farming not well adapted to the crops grown, with poor practices to control runoff, and with inadequate fertilization for vigorous plants. The hazard of erosion can be high in areas subject to drought that weakens the plants and exposes the surface to sharp heavy beating rains that stir the soil and allow little time for the water to percolate down into the soil. The erosion hazard is also greater on soils that are silty in the surface with claypans or other slowly permeable horizons beneath than on soils pervious to water. In extreme instances without control measures, the exposed wet surface silty soil simply flows down to the lower slopes and may fill waterways and ditches.

In areas with few periods of drought or sharp heavy showers, the hazard of soil erosion is less, as in much of northwestern Europe. In many tropical areas the soils are highly pervious. Even with very high amounts of clay, if the clay minerals are not the kinds that swell when wet, the erosion hazard is low even on strong slopes since the water moves down beneath the surface (Figure 50). Thus the hazard of erosion is a matter of the local kind of soil, the characteristics of the local climate, the density of the vegetation, and measures for runoff control.

The amount of sediment resulting from the erosion of soils used for farming thus varies widely. In some catchment basins or watersheds a great deal may accumulate from poorly managed fields over that normally accumulated under the native vegetation. The sediment may come from a large proportion of the catchment basin.

Ironically, some of the most dramatic erosion features—gullies, rocky pavements, and accumulations of sediment—are in some deserts. There is too little water to support plants, yet a very hard shower hits every 2 to 5 years.

Thus in some streams, considerable sediment originates from farm lands. The bulk of the sediment coming into some streams, however, does not come from farm land. It comes from deep gullies and from the banks of winding streams. In many catchment basins, the sediment comes from only about 5 to 10% of the area with these critical sources (Figure 63). Such active sources of sediment threaten the lives of dams built to store water or for hydroelectric power or other uses. Many high mountainous areas have naturally severe torrents after snow melting that are very difficult to control. On reasonably well managed farms, the erosion simply moves the soil material from the upper part of fields to the lower parts.

There are, of course, gullies on farms, caused by concentrated water from bad designs of roadways, misplaced culverts, poor terrace outlets, and the like. Then too, considerable sediment comes from roads and other construction sites. If much of the soil is left bare erosion can be very high. And if much of an area is covered with pavements and buildings runoff may be increased enormously. On sloping potentially erosive soils, the bad location of forest roads and skid trails can cause serious slips, landslides, and gullies. And so can overgrazing of hilly pastures that are left with weak cover.

As climates become drier and the vegetation weaker, erosion usually increases. A sloping infertile soil that is cropped is almost certain to erode because of inadequate cover and roots to hold the water and soil. On the other hand, in humid regions some natural erosion is helpful. Each year a small part of the very surface is removed. If the total thickness of the whole soil remains about the same, as in nature, new fresh minerals are incorporated into the lower part from the earthy parent material beneath. In fact, nearly level soils subject to strong leaching tend to degrade in productivity if fresh materials are not added either to the top or the bottom of the soil. Even so, the nearly level soils lacking hardpans may be highly responsive to fertilizers and other soil amendments.

In farming, gardening, forestry, grazing, construction, and

FIGURE 63

Gullies in a dry area with soft rocks are a great source of sediment in streams.

FIGURE 64

In clearing for pasture in this area of New Zealand, cultivators left strong shrubs on the critical places to control water and erosion.

FIGURE 65

Steeply sloping soils are terraced for tree planting in Algeria.

most other uses, erosion control results basically from runoff control, which we have already discussed. Plant cover plays a great role. I repeat that on many potentially erosive soils it is vital to select well-adapted crops and maintain adequate supplies of nutrients. In the middle latitudes this means also growing cover crops or maintaining crop residues during rainy winter months. The requirements for erosion control in different combinations of kinds of soil and farming and forestry systems are legion (Figures 64 and 65).

As we have seen, the hazard of erosion damage is greatest on soils with low permeability to water due to hardpans, high proportions of clay minerals that swell when wet, and shallowness over rock. Thus there is no direct relationship between soil slope alone and soil erosion hazard. The black soils with large amounts of swelling clays (Vertisols) when used for cultivated crops without terraces can suffer serious erosion with only 2% slope, while some excellent old tropical soils (Oxic-Inceptisols or Tropepts) are so permeable that they have little hazard from erosion even up to 40% slope unless the running water is concentrated on a small part of the field (Figure 50).

In many parts of the world, especially in the rainy tropics, cultivators use mixed cultures of trees and crops, partly to avoid erosion and partly to have the nutrients brought up by the trees, as in shifting cultivation (Figures 66 and 67). Several examples for different sites are shown in FAO Agricultural Development paper No. 81 (1965). In the paper there is a confusion between slips and erosion, which result from different processes, although they may be related. Yet basically erosion control is a matter of runoff control, adapted crops, and an adequate balanced supple of plant nutrients, including plants used primarily for protection of terraces and other structural earthen works.

SOIL BLOWING CONTROL

Like the erosion and deposition of earthy material by water, its movement by the wind has been an important process of soil development. Some of the best soils of the world are developed from

FIGURE 66

This is an example of mixed cropping with Manila hemp (*Musa textilis*) in gallery forest of *Terminalia superba* in the western part of the Republic of Zaire.

FIGURE 67

A home in the midst of mixed cropping under coconut palms is pictured in Kerala State, India.

wind-blown silt that was originally laid down in the wide braided streams from glaciers. I have two photographs of the process, one of the fine earth blowing from a glacier-fed stream in Alaska, and another of a similar stream with little wind in the South Island of New Zealand (Figures 18 and 19).

Actually many soils have had small but frequent additions of dust. Perhaps most of this came from desert areas and other areas that are subject to extreme drought from time to time. Volcanic ash also gets blown by the wind. Yet during drought, many soils without cover contribute dust. An excellent classical study of the process was made years ago by E. E. Free (1911).

Locally the process of soil blowing is widespread even in humid regions during drought on soils recently prepared for crops and left nearly barren. (Many authors use the term *wind erosion* for what is called here by the older term *soil blowing*.) It seems better to limit *wind erosion* to the sand blast on rocks, because the causes, effects, and measures of control of soil erosion and of soil blowing are different.) As with soil erosion, a thick plant cover controls it. And of course sand dunes near the sea blow a great deal unless carefully planted and fertilized. Yet some soils blow badly mainly because of very low supplies of one or more plant nutrients. Especially extreme deficiencies of phosphorus may prevent the coming in of plants.

One may expect soil blowing in deserts at any time the winds are high and in the margins of deserts during drought. Commonly, the blowing continues until the fine materials are gone and a layer of coarse rock fragments or cobbles is left on the surface, called a desert pavement (Figures 68 and 69).

For control near deserts where the soil is dry much of the time, reseeding in favorable seasons and strict grazing control is necessary. In areas suitable for cropping but commonly plagued with droughts, crops and grass can be planted in strips at right angles to the main wind currents. Some farmers in the Soviet Union (and elsewhere) plant rows of sunflowers, nonweed mustard, and other medium to tall plants, partly for protection and partly to hold the snow to melt on the grain fields (Figure 70).

Shelterbelts of adapted trees or tall shrubs have been used in many parts of the world to reduce the effects of either or both cold and hot winds, and to reduce the direct force of the winds on the soil surface (Figure 71).

In the earlier shelterbelts near grain fields, many cultivators planted the trees too thickly and even bordered them with shrubs.

FIGURE 68

This is characteristic desert pavement from blowing of soil material, southern California, USA.

FIGURE 69

This is a desert pavement in southern Israel. Note the mounds where it seems likely that olive trees (*Olea europaea*), mulched with small stones, were grown, hand irrigated, and abandoned centuries ago. As the winds blew out the fines, mounds were left to mark where the trees had been..

FIGURE 70

Sunflowers (*Helianthus annus*) are grown for protection of wheat in the "New Lands" near Shortandy, USSR. Courtesy Shortandy Experimental Station.

FIGURE 71

A shelterbelt of oak (*Quercus* spp.), ash (*Fraxinus*), elm (*Ulmus laevis*), and other trees for greenbelt is located near Volgograd, USSR. (See Figure 52 for preparation.)

Such a tight shelterbelt reduces the effect by causing strong turbulent air on the lee side. By reducing the tightness near the bottom of the shelterbelt to allow some air to pass through, the soils and crops have protection for a greater distance from the belt and more snow is left on the fields in winter. Then too, if belts in grain fields are thicker and more closely spaced than necessary, some area is wasted as well as the soil moisture used by the trees. For the protection of fruits, houses, and barnyards, the belts do need to be thicker and more closely spaced than in grain fields.

In droughty areas it is commonly difficult to establish trees on clayey soils with swelling clays. Trees on these have difficulty getting enough water during drought in contrast to more sandy or loamy soils. In dry areas young trees, at least, may need to be irrigated. As the belt ages it may be necessary to thin the field shelterbelts. It is also helpful to keep out weeds and grass and perhaps trash that has blown in to avoid accidental fires. In many places where shelterbelts are used in helping to control soil blowing, one should not expect complete protection, especially after three or four very dry years.

In any specific part of the world, advisors on shelterbelts need to be resourceful in learning from local cultivators and others what trees and shrubs can be expected to do well. It should be added, however, that within tree species there can be considerable range in hardiness depending on the seed source. Several examples of the use of windbreaks and shelterbelts in various parts of the United States are illustrated in a bulletin by Arthur E. Ferber (1969).

CROPPING SYSTEMS AND VARIETIES

Certainly many thousands of cropping systems are used successfully in the world. They vary, of course, on different kinds of soil and according to the size and potential productivity of the holding. Good ones vary according to the proportion of food products for home or village use, the proportion needed for animal feeding, and the proportion of industrial crops for sale, including forest products. They depend on the skill of the cultivators and the services available to them, including markets and terms of trade.

On many kinds of soil it is useful for several reasons to change the kinds of crop grown rather than plant the same crop year after year in the same field. On soils that have clayey hardpans beneath, the continued growing of crops with shallow roots can lead to trouble. But if a deeply rooted crop is grown in rotation, the lower soil can be kept more pervious. By using rotations of crops diseases, weeds, and insects may be reduced. At one time crop rotations were emphasized more highly as a general rule to follow than they are today. Much again depends on the skills of the cultivator, his equipment, use of fertilizers, and his potential markets.

Many farms are very small and poorly served; others are highly capitalized under individual or corporate management. Some are served in various ways by strong cooperatives. In the Soviet Union both collective and state farms are operated by groups of people working together, and so on. Thus only basic principles are dealt with in this chapter.

Generally we can say that changes from shifting cultivation and mixed cultures by hand labor to the more familiar "open field" farming of the textbook is a slow process. Rapid changes in skill and in village life are not usually practicable. Yet well-educated farmers with reasonable funds have made rapid changes because they already had many skills and, commonly, the assistance of a good advisory service.

As emphasized earlier, some poor cultivators have been urged to change a number of practices at the same time, such as those for water control, fertilizer use, selection of varieties, and pest control. Others more primitive, would be hopelessly confused by trying to learn so much at once. Perhaps the introduction of a marketable, adapted cash crop would be the best start. Yet this does not mean that progress must be so slow as that of our early ancestors. We have learned not only a great deal about farming but also much about working with and advising cultivators.

Control of pests

Pest control, including diseases and insects that influence both crops and animals, and weeds, will play an especially important role as new systems are introduced. Many of these controls require collective action and assistance from government. A switch from mixed cultures or shifting cultivation to open fields can lead to greatly increased weed problems. New varieties grown successfully on similar soils and with comparable day lengths elsewhere may be attacked by known or unknown diseases or insects to which they are not resistant. Such hazards of diseases and insects are especially great if the newly introduced variety is so superior in yield that most cultivators adopt it so that if a new disease or insect comes it can spread rapidly from one field to the next and cause a near disaster in a large area.

Multiple cropping

Two, three, or four crops per year on the same hectares offers opportunities for increased production per hectare in tropical, subtropical, and warm temperate regions. Here we are referring to field crops or to mixed cultures of several crops. Of course, some double cropping is carried on in temperate areas under favorable conditions.

Much depends on the amount and distribution of the rainfall, the humidity of the air, and the frost-free period. Air humidity can make a vast difference between areas having similar seasonal rainfall and temperature patterns. Because of the direction and source of winds, elevation, and so on prospective areas for cropping only a few kilometers apart can vary widely.

We have already explained how contour terraces for runoff control can cause extra water to be stored in the soil for a crop during the dry season. Where water of good quality is available for irrigation, more than one crop can be grown in the dry season.

We need to recall, however, that dry seasons in warm areas as well as winter cold in temperate areas, have "therapeutic" effects on weeds, insects, diseases, and other pests to people and crops. Thus generally speaking, multiple cropping in warm countries may raise new problems for the cultivator. Commonly this means new skills and somewhat more expense.

Yet under good conditions the additional crops per year can give much more return for both land and labor. And different kinds of crops can be grown to improve the local diet. In this work at the International Rice Research Institute, Bradfield (1969) has shown that good systems for multiple cropping can be designed and taught. His research results are suggestive for many tropical areas. With a good advisory program we can expect increases in multiple cropping, especially in highly populated farming areas where land is scarce. Such developments may bring demands for land reform and for new settlements on unused soils that could be used effectively. Reviews of experience and some of the available estimates of results and increases in multiple cropping have been summarized by Dana G. Dalrymple (1971). But, of course, reliable data by countries are hard to come by in most parts of the world.

As with other farming systems that require several inputs in combination, progress can be expected to be slow unless cultivators have incentives, credit, and considerable guidance to get started, backed up by competent research. We must warn, however, that multiple cropping, like new cereal seeds, fertilizer, irrigation, or any single practice is not a panacea.

4

Improvement of farming

In this chapter, we shall look broadly at several aspects of farm improvement. None is discussed in great detail but none should be neglected in the combinations of situations and actions that interact to make a successful agricultural improvement scheme for an area.

Generally the improvement of farming in undeveloped areas means changes from low yields, mainly labor intensive practices, and low, uncertain incomes, to the use of more industrial inputs, higher yields, better terms of trade, and greater security of income. It also means dealing with several technologies in combination as they fit the local kinds of soil. We repeat, these technologies cannot be introduced all at once in the more primitive farming areas.

Planning in such areas is handicapped by inadequate statistics about hectares of farm production, yields, potentially arable hectares, food supply, population, and the like. There is a lack of specific scientific data, including adequate soil surveys for planning farms and good market towns, and a lack of good roads.

What can be done depends also on existing transport, marketing facilities and related infrastructure, local laws, and especially local leadership. Although many things appear to need doing at once, priorities must be made with the help of the people themselves. Rarely can planning be done on a "clean slate." People have their own lives and interests; they have at least somewhat different goals and different values as individuals, as families, and as local groups. Actions that might help most could injure others.

For example, the introduction of new tools and other products may deprive local blacksmiths, cart makers, weavers, and others of their livelihoods. Land holdings may have to be rearranged with fairness. Too much speed can result in making progress for the families with large holdings and capital at the expense of the others. Too little speed brings discouragement.

The job of those giving technical assistance must be guided by the principle that the late Prime Minister Nehru explained to me in 1958: "We must achieve justice but only by just means."

INCENTIVES

With incentives for change most societies in the less developed countries become more flexible. Without them, people tend to retain their traditional methods. With incentives, however, the cultivators purchase inputs, such as fertilizers, tools, and the like. Gradually, of course, these may reduce farm employment and jobs need to be created elsewhere in the trade area, especially in the nonfarm sectors of agriculture.

In many undeveloped areas farm prices have been held down to supply cheap food for the cities. The same result has been accomplished by the sale of industrial and other crops through some governmental "marketing boards" that offer the *only* market to cultivators. These agencies pay cultivators considerably less than the world price but sell at the world price. This gives money to the government (except that some of the profit may find its way into political purses) which amounts to an unfair tax on cultivators. The fact that the prices to cultivators may be as high or higher than those paid by private markets is an unacceptable excuse.

Such practices lead to higher urban incomes, for those with jobs, than to higher farm incomes. The inequalities encourage migration to the old cities where actually many people may already be unemployed. Jobs are needed nearby. With assistance and enlightened local leadership, cooperatives can be organized in some areas that give the cultivators better terms of trade.

Land tenure

Many cultivators suffer from low incomes and insecurity because they do not own the land they cultivate. Some are landless laborers, share-croppers, or insecure tenants on large holdings. In Europe during the end of the Roman Empire and for a long period thereafter land was held in large feudal estates. To a considerable extent this system was followed in Asia and passed to the Americas, especially to Latin America, and later to certain parts of Africa during the early colonial period.

Still in parts of these countries, especially in Africa, land has been held by tribes, lineages, clans, or families. For some time primitive cultivators did not understand land ownership in fee simple. They regarded land as a "free good," like the air, sunlight, or rain. Every cultivator-family in the group was entitled to a suitable plot for growing food and to grazing rights. This even included people who had left for a time and later returned. (A similar understanding today might be that each person is entitled to a paying job.) But no one from the outside—no "stranger"—could have land. In some places, however, these rights were taken away, or were purchased for plantations and estates on which the people were employed without approval of the local group. The old Belgian Congo, after World War I had regulations to prevent this.

At different times and places attempts have been made to expropriate the large holdings, with either outright confiscation or with compensation, and allot parcels to peasant families. Many of the large holdings were not used so intensively as they could have been. For example, some excellent soils for maize, wheat, and other crops were used only for extensive grazing. But the owners would not sell. Examples can be seen today of such under-use, even in areas needing more production and more jobs.

In different parts of the world many schemes have been tested successfully for giving cultivators more direct interest in the land they cultivate. In eastern Europe and parts of Asia, collective or state-managed farms have been organized to permit the use of large equipment and better practices. Yet, even on some of these too many workers were left on the farms after they were mechanized.

Cooperatives made up largely of peasants with small farms have been highly successful, especially in the Scandinavian countries. Much of this sharing of ownership with the peasant farmers has been brought about by various schemes of "land reform" or "agrarian reform." These terms are used for both revolutionary schemes and governmental schemes with compensation. However, the cultivators must be reasonably well organized and determined, and must obtain technical and financial assistance from the government, including an effective advisory service, if these schemes are

to achieve both social justice and high production with better incomes and greater security.

The Food and Agriculture Organization (FAO, 1970, 1971)
has sponsored studies of land reform in recent years that are helpful for gaining a view of the general problems but are weak on
techniques, except for Agrawal (1970). The revolutionary view-
point is set forth in application to Latin America by Frank (1969).
The anthropological background of the political problems is discussed usefully by Andreski for Latin America (1966) and for
those parts of Africa that he has studied (1968).

Unfortunately many sorts of rearrangements have been included under the terms *land reform* and *agrarian reform*, including
the development of entirely new land cleared from forest or made
arable through irrigation and related water-control schemes. For
the subdivision of large holdings to give cultivators more direct
access to land they can control and from which they get full advantage of their labor and management, the term *land reform* is
best. The "consolidation of fragmented holdings" is quite a different problem, although very important in many places. The term
agrarian reform is far too vague to suggest any specific proposals
among a great many to improve the status of cultivators and their
services, including those discussed in this and other chapters. In
this chapter we shall be concerned primarily with the problem of
improvement of existing farming areas and in the next chapter
with the development of new lands. Still these categories overlap.

The subdivision of large holdings under land reform

Land reform programs present many political problems, and are
not without difficult technical problems too, yet social justice for
peasant farmers may demand them. The need for land reform is
urgent in several of the less developed countries and even in parts
of at least some of the developed ones. Large areas in Asia, Africa,
and Latin America have this problem. As Solon Barraclough
(1969) has written, "In poor agrarian societies, land is the main
source of wealth. As a result, the control over land largely de-
termines income, wealth, and power in backward agricultural
areas." He goes on to point out the enormous need and urgency
of land reform. Despite the need, neither the agencies within the
United Nations group, nor the bilateral technical assistance pro-

grams have really tackled much of the problem substantially; nor have many local governments.

In the same issue of *Ceres* (FAO) for November-December 1969 in which Barraclough made his plea, Ernest Feder and others give examples that need not be repeated here. Anyone traveling in the rural parts of these areas can see for themselves. Others deeply interested can read that issue of *Ceres.*

Each situation is unique because laws governing land titles and compensation for expropriation vary between countries and between states or provinces within countries. Great differences exist also in the tribal or common law. In Africa, for example, one may find today nearly every transition from old tribal law to the common law of the former colonial powers in Europe. By understanding the local legal situation much trouble can be avoided.

Where the older estates of latifundia were well managed, the first result of changing the pattern to peasant holdings may well be reduced production at least for a while. Although most of the cultivators may have reasonably good skills for work in the fields many have limited skill in farm management, in the handling of money for purchasing proper inputs—machines, chemicals, and other supplies— and for marketing their products. Thus success nearly always depends on a good advisory staff to help them get started toward good management.

To achieve a fair scheme for the cultivators, a *good soil survey is an immediate necessity* to avoid both failures and corruption. Many failures could have been avoided with them. As we have emphasized in Chapter II, the soils need to be evaluated according to their potential response to management. Kinds of soil that may look alike to many people actually vary greatly in their potential under foreseeable farming systems.

It is *highly irresponsible* to divide land arbitrarily into blocks of 10, 20, or 30 hectares, for example, without considering their potential productivity per hectare. A scale of productivity for the expected crops can be worked out for each kind of soil, and a productivity index established so that blocks can be reasonably alike in actual potential, considering both the potential responsiveness of the soils to management and the size in hectares. As a common standard one block might need to be 20 hectares and others 30, 40, or more hectares. Some parts may need to be allocated to forest use or extensive grazing.

In schemes for irrigation these problems of fair distribution are even greater. The laws governing water rights may be even more complex than those governing land rights, and soil response is more variable because of differences in water quality, hazards of salinity, and in drainage requirements. I have even read of proposals in projects with one standard size and value for unirrigated land and another standard size for irrigated land. Unless the pattern of soils is unusually simple, this can be completely irresponsible, because different kinds of soil, even among those that "look alike," can vary widely in response to management under irrigation.

With such errors land reform schemes break down. Without a good plan based on a study of the soils, water, and essential services, great delays and failures are inevitable. For success a good plan should be made on the basis of an accurate soil survey and cadastral survey, and the scheme pushed through rapidly. Many have cited as an example the now famous Gezira Scheme in the Sudan on difficult kinds of soil (Vertisols, or Tropical Black Clays). Yet they may not know that the success was preceded by years of highly competent research in soils, in soil and water management, and of the problems of crop varieties, diseases, insects, water control, and so on (Tothill, 1948). Without this research, not directly applicable to other kinds of soil, the scheme would have failed miserably.

A graduated land tax has also been used in several countries to break up excessively large holdings, especially where peasants needed land and where land in the large holdings was not effectively used. Examples include large holdings where all or parts of the soils were used for extensive forestry or grazing but which would be responsive to much more intensive use. For farmers' holdings the tax rate should be based on both area and potential productivity, not hectares alone. Thus the law may provide that the first 10 or 20 hectares, as adjusted for potential productivity, is tax free or is taxed at a small standard rate per hectare, say x. Then other blocks in holdings of 20 or 40 hectares, as adjusted for potential productivity, are taxed at $2x$ per hectare, and so on through $3x$, $4x$, and $5x$ up to the size of the holding as adjusted to the potential productivity. If the principal need is to have the better soils for cropping for peasants, the other nonarable land could be ignored. Yet as it turned out, the failure to include the land used only for forestry in the land reform scheme in Japan was probably a serious error. Some of this land was left in large private holdings without provisions for good forest management.

Consolidation of fragmented holdings

This work is very badly needed in many farming areas for effective water control and soil management. Although quite a different kind of problem, it is unfortunately lumped by some with "land reform." In some areas perhaps elements of breaking up large holdings or the development of new land should go along as part of the scheme.

In many old farming areas, the village lands have been fragmented through inheritance, marriage dowries and other transfers, and by canal and road construction, until a peasant holding of 3 or 20 hectares may be in 5 to 15 unconnected fragments. Effective systems of management that include terracing for runoff control, irrigation, drainage, crop rotations with pasture, fencing, or the use of machines are nearly impossible. Obviously very much unnecessary work is required for a unit of production. I have seen peasants leading irrigation water around other plots for one-half kilometer or more to irrigate one-quarter of a hectare. Then too, most of the materials must be carried by people to and from the plots.

This was a big problem in Europe. But now most European countries have already carried out programs for consolidation of holdings (Binns, 1950) and have laws intended to avoid the splitting of existing farm units into sizes too small to be viable as farms.

Increasing efforts are being made in the less developed countries to correct the problems of fragmentation, but they are not yet adequate. In many places it is a first essential step toward farm improvement.

Despite the enormous potential benefits, the consolidation of fragmented holdings should not be undertaken lightly. With potential for intensive farming, an accurate detailed soil survey is essential. Where possible this should be supplemented with topographic surveys. These can now be made easily by using modern plotting devices with stereo pairs of good quality aerial photographs. An accurate cadastral survey or plotting is needed at the beginning and at the end of the project.

In many highly populated areas, it is not practical to follow the usual goal in Europe of getting all the land into only single holdings. Many village lands include two, three, or more unlike kinds of soil with the sets of practices most likely to be used by cultivators. As pointed out earlier in the discussion of the subdivision of large holdings, *total* hectares is *not* a fair guide. The hectares belonging to a cultivator must be weighted, fragment by fragment, up or down, in reference to a standard for potential productivity in the village so that when the scheme is completed, each cultivator has a production potential similar to what he had before but not necessarily the same hectares or the same total area.

Nor can all land for a family always be placed in one block. Where the kinds of soil are contrasting they may be placed in two or three groups and each family may get two or even three plots. Some soils may be useful only for forage production, for example, and it would be a great injustice for one family to have only that kind of soil.

Reconsolidation is especially important where water control is vital, either by controlling runoff, by irrigation, by drainage, or by some combination of two or even all three of these practices. With rain water controlled to soak into the soil, scores of millions of hectares in India alone could give much higher yields. I have seen an added benefit in some villages; the ground water was increased so that enough water was available for irrigating more garden plots. For such control on gently sloping soils earthen terraces *must* be laid out at the *proper* angle to the contours. They can rarely be laid out on old property lines without causing ponding or erosion, or both. On hilly soils, stone terraces may be needed. With the terraces accurately designed for the kind of soil and local rainfall pattern, and properly laid out, local labor can be used to move the earth with bullocks or to carry it on the head. Machines are not necessary. Where labor is abundant this is a proper labor-intensive program to use the funds for jobs within the village area.

An individual, or team leader, carrying on this work must be highly competent. Not only that, the cultivators must trust both his competence and his integrity. He must become acquainted with them and their problems and goals. After all, their livelihoods are at stake and they are the ones who will live there. When poorly conceived schemes have been attempted and have led to bad results, the village cultivators feel they have been cheated. Then the cultivators in other villages resist change or become extremely cautious.

Yet good schemes have been carried out successfully. One of the good ones I have seen was developed for Nawdika Village near Hazaribagh in India (Kellogg and Orvedal, 1969). Since the village lands included two highly contrasting kinds of soil, the fragments in these two parts were consolidated separately (Figures 72 and 73). On the better soils the new property lines are along the terraces and at right angles between them.

FIGURE 72

Small parcels of good soil are consolidated after proper terracing for water control in Nawdika Village near Hazaribagh, India. A man stands in the grassed waterway that takes away the excess water. Yields were doubled and can be increased much more with fertilizers and crop varieties that respond to increased nutrients and moisture.

FIGURE 73

The poor soils of Nawdika Village, suitable only for grasses and fruit trees, were consolidated separately. This photo shows the outlet for grassed waterways.

One part is suitable only for hand-harvested grass and widely spaced fruit trees. The other part is suitable for intensive cropping with terraces for runoff control, to give the crops more moisture during the rainy period, and enough is stored for a second crop during the drier period. This alone doubled production in the village. Then with fertilizers, improved varieties with the genetic potential to respond to more plant nutrients and water, pest control, and related practices, yields can be seven or eight times what they were before consolidation.

In many places with small fragmented holdings, schemes for consolidation may be complicated by problems of land tenure in achieving both effective soil use and fairness to all village families. Thus in programs for the consolidation of fragmented holdings, as in land reform, equity of taxation, fair tenant-landlord contracts, and credit may also be needed for good results. Where nearby large holdings can be subdivided, these areas may be added to the general pool by expropriation or purchase. Near a few villages some unused land may be reclaimed for the same purpose. Yet in many areas "land reform" in the usual sense of breaking up large holdings may not need to be an important factor.

In fact the consolidation of fragmented holdings is such a different problem than what most consider to be "land reform" that special laws may be needed to carry it out. Such laws must be thoroughly understood by the cultivators. Preferably the laws should be as nearly local, say within states or provinces, as the framework of government allows. Grants from the national government may be needed to get the work done if the local governments lack staff and adequate sources of taxes.

I am enthusiastic about the great potential of such schemes in millions of villages in the world where traditions of private farms are strong. Success also depends on a good advisory program to take advantage of the increased potentials. Good storage facilities and marketing outlets must also be arranged as well as some credit to help get started with the improved practices.

Yet advisors of poor peasants and managers of such schemes are warned against large machine operations where unused labor is abundant. It is foolish to buy or rent such machines and have crowds of poor unemployed villagers just standing around watching them operate. They are rarely necessary and the money for their hire or purchase does not remain in the village.

But no aspect of technical assistance requires greater skill in soil science, the engineering of water control, and the economics of land use than this kind of local planning. And certainly none that I know of requires greater command of those arts of communication that lead to confidence.

I have seen many other valuable side effects of success in this work. As villagers learn that they *can* cooperate effectively, they have also established schools, storage facilities, and other needed village services that they had formerly thought they could not do by themselves.

Where peasants have already a strong cooperative and communal background of sharing work and production, the consolidation of units into manageable fields is simpler. But the basic principles of soil and water management are the same. This consolidation of fragmented holdings is emphasized so much here because of the great opportunities for improvement in local areas, with generally less cost and political opposition than the splitting of large land areas held by politically powerful people.

Farm services

Many cultivators in parts of both the developed and less developed countries are held back by the lack of either private or public servcies, or both. This is true for many on highly responsive soils that are far from a good town or even useful roads. Unhappily some cultivators have taken up land in areas not served by good highways, boats, or railways. Many are cut off from the mainstream of trade and commerce by unbridged rivers, high mountains, dense forests, or other obstacles.

All inputs, such as fertilizers, pesticides, machines, petroleum products, electric power and the like are very expensive or even nearly impossible to have. Storage space for crops and processing facilities are equally inadequate. A first step toward rewarding farming in such areas is serviceable transport and a market town if the cultivators are to rise much above the level of bare subsistence and are to make reasonable contributions to their needs and the needs of the world for food and other farm products.

Even some of the small countries, such as a few in Africa, have such a small population that the market within the country is too small to permit viable manufacturing concerns or even efficient service agencies unless fishing, forestry, or mining can be further developed to share the costs of such services with farming.

The lack of planning, or of opportunity for it, has blighted more than one settlement scheme in isolated places. The area of excellent soils drained by the Peace River of northern Alberta in Canada was near the edge when I saw it in 1951. Later, petroleum was discovered which gave an opportunity for a sharing of the costs of the infrastructure with farming and forestry. Many of the services suggested in this section have been covered usefully by A. T. Mosher (1966) and in a report of the Ford Foundation to the Government of India (1959).

With the consolidation of fragmented holdings and with more forage, many areas have great potentials for more livestock production. As we have already mentioned, disease control is an urgent matter and provisions need to be made for specialists in this area, both in educational efforts and the advisory service (FAO, 1963).

Education

Some farming areas lack schools for the children of cultivators so they may be able to read the useful pamphlets and books that are available in a written language. A successful cultivator should be able to read what is written on the fertilizer bags, the directions on machines and tools, and many other items in order to work out the best system for his holdings. To avoid having to wait for a generation, early progress requires a vigorous program of adult education. Much of the good progress on farms in the Soviet Union, for example, is said by many to have been due to a massive program of adult education after World War I and still continuing (Myrdal, 1968, 1971).

During the past 20 years, perhaps too much emphasis has been given to the two ends of the educational ladder in the less-developed countries: (i) on primary education for the national prestige of having a high percentage of literacy and (ii) on higher education for the personal prestige of a few professionals. The area between— the high school and junior college—has had too little emphasis in many of the newly developing countries. Besides training in the basic principles of farming, courses are needed for potential mechanics, electricians, and other trained people required in agricultural development.

Yet a great handicap in agricultural education has been old textbooks and old ideas of former systems of farming, or even old systems in other countries. As new countries develop from now on nothing will change faster than their agricultural systems. These changes are not only in farming but in the whole field of agriculture. Much emphasis on the simply "vocational" is wasted time. The student in agriculture needs the basic skills in mathematics, science, and language. Much that he can use later will come from reading and conversations. But he needs to know the basic words and concepts to understand. Some of the training is too exclusively specialized. There is no harm in specialization, in fact we need more of it, but *not* at the expense of basic education.

We should avoid the extremes of the narrow specialist and of the shallow generalist, who has a little of "this and that" but no exposure in enough depth in any field to understand the principles of scholarship, which are basic to continuing education. Probably the shallow generalist is less useful than the well-trained but narrow specialist. What is most helpful is the *educated* specialist who can appreciate scholarship in other fields important to agricultural development.

This subject has been well covered by Philip H. Coombs (1971) in his paper "Education on a Treadmill" in *Ceres*. In the same issue of *Ceres* Andrew Pearse puts forward some examples of great needs for educational reform in several countries of South America. (See also Chapters VII and VIII.)

Advisory (or extension) services[1]

Advisory services staffed with competent people who can communicate easily with cultivators in their own language are highly important. Those dealing directly with people in the villages should be able to demonstrate the improved practices. Such village advisory people do not need to be full university graduates. Good texts for a 2-year course can be prepared in their written language. Among the promising candidates are sons of cultivators who can actually get down on their hands and knees, if need be, to show cultivators how to apply fertilizer to a crop, how to transplant seedlings, and all the other jobs that need doing well.

1. The term *extension* seems to have been used by universities in the sense of "extending" their serivces in general education beyond the campus. Yet *extension* may not be the best term for what in agriculture has become primarily in most countries an advisory service to help people, with or without direct connections with universities. (See Kellogg and Knapp, 1966, for an expanded treatment as applied to the United States.)

The good advisor must be able to grasp the principle of inter-actions as applied to farming. He must know, for example, that fertilizers are economically effective only if the correct kinds and amounts are used for each local kind of soil and crop combination, and if the other practices to meet the requirements for good harvests are carried out at the same time. Among primitive people far from places to buy fertilizer, the advisor needs to understand the basic principles of shifting cultivation and the use of manures, compost, and wood ashes rather than commercial fertilizer. Failure to grasp this important principle of interactions is the greatest technical handicap to the improvement of farming or to the establishment of farms on new lands (Kellogg, 1962–1963).

Such failures can be illustrated with a statistical study of the affects of the use of fertilizers published by FAO in 1962 (Williams and Couston, 1962). Curves presented on pages 7 and 15 show that in some countries fertilizers were used along with other good practices as needed while in others they were used only by themselves. The yields were enormously different. Perhaps unintentionally, the results dramatically illustrate the need for a balanced advisory service for farming. Rarely does improving one practice alone do the job.

As pointed out by Gordon (1969) writing about West Africa, cultivators show considerable skill in adjusting to the local kinds of soil and to local risk factors but not in combination of practices to alter the soil for increasing production. And there is available plenty of labor to improve roads, water supplies, and storage facilities. The advisory officer thus has the job of helping to solve sociological problems for the welfare of the whole area where he works, not just for the cultivators alone.

A successful advisory service has close working relationships with research and teaching centers. In fact, much of the research and demonstration work should be cooperative between the research centers and the specialists in the higher levels of the advisory service. This is an advantage of the colleges of agriculture in the land-grant univerisites of the United States. In making a studv of these colleges (Kellogg and Knapp, 1966), we found that the best advisory (extension) services were in those institutions that had strong research and strong teaching programs. In fact, if any one of the three—research, extension, and teaching—seemed to be weak, usually the other two were also weak.

The degree of excellence in each goes back to the dedication within the institution for cooperation, scholarship, and service to people. And this does not mean dedication to farmers only but also to those in the other sectors of agriculture that are essential to a viable agricultural system that serves both farmers and the users of farm products. In fact, many of the greatest opportunities for increasing the world food supply and the incomes of farmers are reductions in wastes between the fields of standing crops and the kitchens.

This illustration is by no means intended to imply that the United States pattern of organization could or should be followed in any other country. But the principles are widely applicable within those social systems that I know.

It would not be difficult for a highly competent advisory officer to go to villages with soils and potential farming systems familiar to him and to help cultivators increase their yields very greatly. Yet if he overlooked adequate storage, marketing, and processing facilities, local prices could fall so low that the cultivator's incomes could even drop to lower levels than before.

Credit

In the improvement of farming, or for new land development, cultivators need some loans at fair rates of interest. Without the benefits of credit the new combinations of practices could be limited to only the more opulent cultivators. Many need credit for investments in fertilizers, water control practices, pesticides, and the new varieties, otherwise they would not be able to take advantage of the newer, more rewarding combinations of practices.

Many kinds of arrangements can be made with a central bank for loans, perhaps to local societies, to supplement capital put into them by the members. (See Ford Foundation Agricultural Team 1959 for arrangements suggested in India as one specific example.)

Since many cultivators have yet to acquire skill in business, some countries have used "supervised credit." An employee of the local, state, or national government needs to examine the plans for improvement by a cultivator, approve the loan for specific purposes, and see his progress from time to time. Such a cultivator can also be assisted by the local advisory officer.

In whatever way credit is arranged it needs to be timely. The money for the purchase of fertilizer, seeds, and equipment must reach the cultivator so that he has the materials in advance of the time for planting. I have seen instances where the loan came so late that the money was used for a dowry or some other purpose unrelated to his farm plan.

The credit arrangements might be handled by a cooperative supported by investments by cultivators with loans from a central bank as suggested above.

Cooperatives

Some kinds of people find it easy to establish cooperatives and to do most of their buying and selling through this device (Figure 74). The Scandinavians somehow have a special talent along these lines. Also cooperatives seem to be easier to operate for certain special products produced widely in the area, such as African oil palm, citrus, cotton, cacao, milk, cattle, and so on.

Advisory officers or agents with considerable skill are necessary to assist cultivators in establishing a cooperative and getting it into good operation. The advantages in both selling and purchasing are large in well-operated cooperatives. In areas with wide

FIGURE 74

A home is shown in a cooperative village in the Republic of Zaire.

ranges in social status among the cultivators, however, the coopera-tive may be run similar to a joint-stock company for the advantage of a few leading members.

Use of machinery

Unhappily some have the notion that all good farming is done with machinery, including tractors and harvestors. Investments are high and considerable mechanical skill is required to keep them in operation. Commonly, cultivators may be urged to buy machines larger than needed or more costly in relation to the wages of hand labor and costs of bullocks or draft horses.

In cooperatives or other local groups one or two people with the capital and skill may buy one or more machines, say threshers, harvesters, and the like, and do custom or hire work for others that need machines for only a few days. Machines can do some jobs so much faster and easier than the older hand methods that their advantage is considerable. Many on small holdings, however, can use the small garden-type tractor to greater advantage than larger tractors. These are widely used by successful farmers in Japan.

Gradually as farming improves and jobs are more plentiful most farming on responsive soils will be mechanized. But the pro-cess should not be encouraged more rapidly than the returns from the investment pay off, especially with costly machines.

Transport

The improvement of farming and the establishment of new arable soils require good transport, from the farm to the local market and from thence to the coastal or railway market towns. In many countries this has been a gradual process since the early days of transport by small and large boats. Then roads and canals were built. These were followed by railways and, more recently, by air service, including freight.

Investments in roads for trucks can be very rewarding. These too can be built most rapidly by great heavy earth-moving machin-ery; yet with good designs and supervisory engineers it may be more economical to do much of the work with local labor, if plentiful, and more of the money can be kept within the area. In the great interiors of Africa and South America, and in parts of other continents,vast areas of responsive soils remain unused, main-ly because of a lack of transport. Canoes penetrated many areas long ago, and airplanes more recently. But for agricultural development railways and truck lines are essential, not only for farming but also for forestry and mining. Many of the streams also have great potentials for hydroelectric power that can be used for improved storage.

I feel certain that at a time to come, perhaps many decades ahead, the area of the Zaire (Congo) River Basin will be one of the most thriving farming areas in the world. The resources are all there. But more towns, railways, highways, and electric power stations will need to be built.

5

New development of potentially arable soils

We have already discussed the opportunities and the principles for increasing farm production on soils already in use through better management systems of farms, improvements in the industrial and marketing sectors, land reform, and in the planning of towns and villages. Now we shall look at the broad potentials for expanding crop production.

A total of about 1.4 billion hectares were being used for farming in the world in 1966. This figure could be increased by adding about 1.8 billion new hectares of potentially arable soils to make a total of about 3.2 billion hectares of arable or potentially arable soils in the world.

The Soil Survey staff of the United States Department of Agriculture has been collecting data on the kinds of soil in the world for many years, including estimates of the unused potentially arable soils. The first estimate for new arable soils (Kellogg, 1950) was only 0.52 billion hectares. The next one (Kellogg, 1964) gave an estimate of about 1.3 billion hectares. The most recent one (Kellogg and Orvedal, 1969) gave an estimate of 1.8 billion hectares of new potentially arable soils not then used for farming. None of these estimates includes desalinization of sea water for irrigation to supplement supplies from rain, streams, or ground water. The higher estimates result from increasing knowledge of soils, especially in sparsely populated areas, and increasing knowledge of their responses to currently known combinations of management practices for both food and industrial crops.

The largest areas of potentially arable soils are in Africa and South America. Africa could increase from about 125 million hectares now cultivated to around 1,600 million. South America could go from around 78 million cultivated hectares to about 648 million. North and Central America could expand by about 367 million hectares. China and Southeast Asia could increase by roughly 80 million hectares and the Soviet Union by slightly over 109 million hectares. In addition Europe, Australia, New Zealand, and the large islands have considerable potential for expansion.

As research goes forward in farming and the other aspects of agriculture, these estimates of potentially arable soils may turn out to be low. Yet bringing forth these facts is in no way intended to suggest an "optimistic" view of the current problems of food and agriculture. Nor do they reduce the urgent needs just now to assist people to have larger harvests and more jobs where they are. Because of accidents of history and migration, the largest areas of potentially good yet unused arable soils are not where the bulk of people needing food and jobs now live.

Yet many do live near these areas and the good arable soils that are needed for use should be developed. Good techniques are now available for mapping kinds of soil and for appraising their responses to management; and for appraising the water resources, and the opportunities for hydroelectric power, mining, and forestry. Also much is known now about the requirements for town-and-country planning to include the essential infrastructure and to avoid crowding and pollution.

Thus in new developments it is possible to avoid some of the most difficult problems that plague people in some older settled areas. Holdings may be made larger and legal provisions provided against undue fragmentation. Areas can be set aside for industries so that air and water pollution are low and far less harmful to people in town or country. This does not mean that all the needed facilities should or could be developed at once, but space can be allocated in advance.

Thus the first steps in the development of new lands are to study the soils, the skill and culture of the people to be settled, and the other potential resources to support an efficient infrastructure for all resource development, including the essential nonfarm sectors of agriculture.

Surveys as a basis for planning and development

Both land reform and settlements of some new lands have been undertaken without soil surveys. But the results show that it is downright irresponsible to do so any more (Agarwal, 1970).

As explained in Chapter II, it is highly desirable to begin the study of a large, poorly known area with a small-scale map of about 1:1,000,000 (1 cm on the map equals 10 km on the ground). An estimated map can be made in advance from available data about the soils and about the climate, vegetation, geology, relief, and the age of landforms. Good air photos are helpful in most places, except with dense tropical rain forest, if samples of the patterns are examined in the field. Once this estimated map is made it can be checked along existing roads and trails. From this map certain areas can be at least tentatively eliminated for farming, although some of them may have value for forestry, grazing, or other uses.

Judgments can be made at this point about the need for a soil map of intermediate scale, say 1:100,000, to decide where more detailed soil surveys are needed. An excellent example is a soil survey of Western Samoa (Wright, 1963).

In dense tropical rain-forest areas, however, air photos are of little help, except to spot streams, roads, and clearings, since the surface of the ground is concealed. For such areas strips need to be cut through the forest for making a useful soil map for finding the good areas. Two examples, with good texts, are one for southwestern Ghana (Ahn, 1961) at a scale of 1:250,000, and one for the southwestern region of the Ivory Coast (Carroll and Malgren, 1967) at a scale of 1:200,000.

After the areas suitable for intensive cropping are selected, detailed soil maps on scales of around 1:20,000 are needed both for soil selection and for good advisory assistance to settlers. An example of a soil survey in a complex area, with adequate text, but with soil maps reduced in scale for publication of 1:63,360, is available for northwestern Ireland (T. Walsh, P. Ryan, and staff, 1969). Areas for development under irrigation and prospective sites for villages and towns may require a scale even larger than 1:20,000 for good planning. The soils along strips for planning precise highway location and designs in areas of considerable soil variation, are commonly made more detailed yet, say about 1:2,500 (Michigan Department of State Highways, 1970).

For soil surveys to be useful they must be interpreted in terms of yields of trees, forage, or adapted field, horticultural, and industrial crops under alternative sets of management practices. Data for these interpretations must come, of course, from the results of experience and research in already developed areas on similar kinds of soil. So the soil names must be given in terms of a standard system of taxonomy so that full use can be made of the results already available from other areas of like kinds of soil in the world. To plan farming systems, other surveys are also needed to appraise the water resources, amount and distribution of rainfall, and variations in temperature and humidity.

Then too, geological surveys should be planned to find opportunities for the exploitation of minerals. Such surveys for phosphate deposits are especially needed in the less developed countries. Both mining and forestry industries can help contribute to the necessary transport facilities and towns required for success with agricultural development. After the specific areas to be used within the general area have been selected, it is necessary to have a cadastral survey with permanent markers for the land lines and a good cadastral map of the whole area.

Planning of new development

As mentioned earlier, to plan a settlement area it is highly important to know the culture and skills of the people to occupy it. If for some reason this cannot be done, no attempt should be made to plan the kinds of houses and the size of tracts, since both will most likely be wrong. One should know the preferences of the people for living in villages or on their individual holdings. Some groups may be prepared at once to buy and sell through cooperatives; others may not. If all such preferences are known in advance the best arrangements can be made (Figure 75).

Size of holdings and unallocated land

A common error in settlement has been to make the individual holdings too small. Even though settlers may lack skills at first and units of about 20 hectares, as adjusted for potential productivity, seem large enough, the chances are that in a short time a unit of that size will turn out to be too small to give rewarding employment to a farm family. With all land allocated then no one can

FIGURE 75

Wheat in the "New Lands" near Shortandy, USSR that will yield about 1.7
tons per hectare.

have a farm of appropriate size until others move out. This sort of
problem arose many times in the settlement of the United States,
especially of the Great Plains during the period, roughly 1860 to
1910. The error was repeated in Alaska in about 1935 to 1942.
This problem may be partly avoided by leaving areas of unalloted
land that can be added to farms needing more land later. If it turns
out that this land is not needed, it can be alloted to other uses
later.

    Costs for settlement vary widely depending on what prepara-
tion beyond the basic surveys already discussed are required
(Agrawal, 1970). Then some costs may be required to study the
people wanting land and for selection of the settlers. Both the
settlers and the advisory staff may need training.

    Many of the most promising soils for new farming are covered
with forest, as were most of the good soils for farming in the northern and western parts of Europe and the eastern parts of North
America, and will need clearing. Parts of allotments may be left
uncleared at first, except for the huge trees near the margins that
might be dangerous. After all in the tropical rain forest the trees
help to hold one another from falling.

    Essentially all the items discussed in the previous chapter on
improving existing farming areas must be given attention such as

arrangements for credit, marketing, machinery, cooperatives, and transport.

Many forms of organization have been successful, depending on the people and what they want, including the kibbutzim of Israel and collective farms in the Soviet Union. Many other people prefer their own private holdings with or without strong cooperatives.

Although the many services should be *planned* at the start so roads, houses, and other buildings are not put in the wrong places, not all must be done at once. On housing, for example, many settlers can build most or even all of their own. If too much is attempted at once, costs will be too high. Yet there is equal danger in excessive delays. Once people get started with enthusiasm for a new opportunity, delays can promote criticism and trouble for the managers (C. Lleras-Restrepo, 1971).

With a good soil survey, including interpretation of estimated yields of kinds of crops and required inputs, estimates can be made of production and income. The major development costs can be borne by the government. Much of the public cost will be returned later in taxes from the greater business, not only from within the project itself but also in the greater business and employment elsewhere in the country. As we have explained, many skilled people besides farm workers are required to build and to maintain the essential infrastructure for agriculture and forestry. So jobs are created for other people too.

Arrangements will need to be made for housing the settlers and the staff, and the usual services—stores, health, schools, water supplies, and the like. A community center will be needed, probably in the central village or town.

The village or town should have around it ample space for necessary houses, and other buildings that will be needed as the project becomes older, including recreational space for the children near their homes. If well located the village may become a service city for a whole trade area. Besides the main roads that service the town for heavy transport, local roads to the farms are needed.

In dry areas all or part of the soils may require irrigation. Wet soils may need drainage. Either of these may require land levelling or smoothing for successful use. If terraces are needed to control runoff, they should be planned for even though not all are made at the start.

In the previous chapter we have already discussed "the subdivision of large holdings under land reform." This term *land reform* properly refers to the subdivision of large holdings into smaller farms, especially for landless farm laborers. Unhappily this term is confused by some with *agrarian* reform, including cooperatives, the consolidation of fragmented holdings, and even new settlements on nonprivate land.

Several studies of specific projects for land settlement are outlined in *Land Reform, Land Settlement, and Cooperatives* (FAO, 1971). In this report several difficulties are highlighted, including (i) wastes because large areas are held privately and owners will neither sell the land nor cultivate it; (ii) poor records of development, yields and costs; and (iii) inadequate legislation. Yet for many of the projects that turned out badly, the greatest error was the lack of a well-interpreted soil survey to have holdings of reasonably fair production potential, and for proper planning of transport, towns, and essential services.

FIGURE 76

Cocoa (*Theobroma cacao*) under partial shade in Ghana as normally grown.

Enterprise selection

Especially in the planning of agricultural development on new lands, in the newly developing countries not only food crops should be included but also other farm products for trade and foreign exchange. Many crops can be grown on responsive soils in the tropics and subtropics that cannot be grown in temperate regions, such as the continental United States, Canada, Europe, and northern and central Asia. For a long time people in these areas have been importing tea, coffee, cocoa, rubber, coconuts, mahogany, African palm oil, sugar from sugarcane, sisal, jute, and other fibers, and so on (Figures 76, 77, 78, and 79). The prospects now of marketing some of these, especially more tea, coffee, and some of the fibers are questionable. Sugarbeets have been used, unfortunately, for sugar in the temperate areas rather than importing sugar from cane (Figure 80). Yet there is evidence that markets in the developed countries may use more African palm oil and rubber, and no doubt other less prominent tropical farm crops.

FIGURE 77

Small young rubber trees (*Hevea*) were planted under shade of bananas in the Republic of Zaire. The scheme is also used for forest tree seedlings. It is called the Taunga system, first used in Burma.

FIGURE 78

A tea garden in India under thin shade is pictured here.

FIGURE 79

New coffee (*Coffea arabica*) is seen just after planting on Tropept soil in Vietnam. The seedlings are protected by coarse grasses for shade.

FIGURE 80

Good sugarcane promising high yields of sugar in Hawaii.

Investments in rubber plantations slowed down with the development of synthetic rubber. But now the outlook is considerably better partly because of higher yielding strains of rubber trees (*Hevea*). Not long ago inquiries of large rubber-using industries in Europe and in the United States showed that about one-third preferred synthetic rubber for their products, one-third preferred natural rubber, and another one-third were indifferent and were influenced mainly by relative prices. Thus rubber is now being replanted. With the new strains grown on suitable soil and expertly managed the market could expand.

Another good crop that seems to be expanding is the improved African oil palm. This crop, too, needs skillful management. Like all oil crops it needs reasonable supplies of available magnesium in the soil for good phosphorus nutrition. When growing on permeable soils it puts down very deep root systems that explore a large area for water and nutrients.

These and other possibilities should be explored in advance so the proper soils, with appropriate climatic conditions, are selected. Also the essential infrastructure for the necessary supplies and for processing and shipment would need to be planned. For example, the oil from African oil palm must be carefully handled to prevent contamination or fermentation.

6

Research and educational systems[1]

Most developed countries have a national system for research and education. As the less developed countries move toward improved agricultural systems they will need both good educational and research programs. Yet there is not, and need not be, a set and common pattern of organization. National and state or provincial governments have a responsibility, and so do universities, both public and private. Business firms that transport and process farm products and firms that manufacture chemicals, machines, and other inputs used by farmers have research responsibilities (Figure 81). Each developing country will need to work out the necessary educational and research programs that fit its need and its people.

Nor can we hope to achieve rewarding farming from farm research alone. Not only is agriculture much broader than farming, but also the people in a country make up a nearly essential market for much of what farmers produce. Unless these people are working effectively at rewarding jobs, the local market may not support modern farming systems for food and industrial crops.

Much of the science and technology so apparent on modern farms today did not arise out of what is commonly called "agricultural" research alone. Many sciences and industrial arts have had large roles. For a well-known example of enormous secondary benefits to farming in a large part of the United States the Tennessee Valley Authority could be mentioned as one of the most suc-

1. A part of this chapter can be supplemented by *Agricultural Sciences for the Developing Nations,* edited by A. H. Moseman (1964).

FIGURE 81

The Broadbalk experimental field at Rothamsted, England, (established by J. Lawes in 1843) is an old and still current symbol of farm research.

cessful large projects for combined resource use in the world (Lilienthal, 1944). It was established as a public corporation during the depression years of the 1930's to harness the Tennessee River and its tributaries for electric power, flood control, and navigation; to develop new and more concentrated fertilizers; and to improve land use. The TVA sought the cooperation of other national and state specialized agencies. These agencies made topographic surveys, soil surveys, and helped in many ways with their regular programs, including research and farm testing of the new fertilizers in combination with other good practices. Such farms were called "test demonstration farms." Possibilities for new crops and livestock were greatly improved with the cheaper and better fertilizers and with the refrigeration made possible by low-cost electric power. It took many skills and much hard work to bring about these great results.

Considerable harm to learning has resulted from the need to classify the many aspects of human knowledge in order to organize libraries and courses for teaching. Many people take these purely arbitrary divisions too seriously. Certainly it would be difficult to teach mathematics, welding, and poetry to the same class of students at the same time. Yet they are related and help to make one whole.

As pointed out earlier, we need specialists in agricultural development; educated specialists with broad points of view about human knowledge, and with enough depth of experience in some one field so that they have learned the tough principles of scholarly research. These kinds of specialists can work together at common problems requiring interdisciplinary effort. The "generalist" without deep scholarly experience is rarely helpful in research. Much of the needed research for agricultural development is necesarily interdisciplinary. The development and testing of an improved crop variety, for example, requires the attention of geneticists, of course, but also of plant pathologists, entomologists, soil scientists, nutritionists, and others. Recently many people have greatly misunderstood some of the developments, partly because of the unhappy slogan "The Green Revolution." Some have said, "These new varieties of cereals with their high yields will tend to deplete the soils of nutrients." Actually the quest was mainly to develop varieties that *would* respond to higher rates of fertilizer application and better systems of water control.

Hybrid maize gave immediate increases in yields on the highly productive soils of the midwestern United States because the older varieties were unable to use all the moisture and nutrients already in the soils. But on the less fertile soils of the Southeastern USA, the new hybrids gave little increase in yield until heavier fertilizer applications were made and the moisture in the soils increased through runoff control or supplemental irrigation. Then too, farmers had to do a better job of soil selection for maize with the new practices than they had before.

I recall an experiment which was designed to compare yields of maize with terraces for runoff control against a similar field without terraces. The "experimentors" even measured the moisture in the soil and found more, of course, where runoff had been controlled. But the yields were similar, so they concluded that terraces were not helpful! Someone who understood the principle of interactions pointed out that to get the benefit of the extra moisture they needed heavier fertilizer applications on the terraced field. This was tried the following year with much higher yields on the terraced field but no higher yields from fertilizer on the unterraced field.

This application of the principle of interactions among the many factors responsible for the final result is simply illustrative of a great many that are now understood. Others remain to be worked out. For example, why does maize not respond to phosphatic fertilizers on a few kinds of tropical soils unless animal manure is also applied? As just a guess, the manure may perform some sort of chelating action. Other tropical soils have still unexplained sources of nitrogen. If the facts were known, management systems might be developed to increase the nitrogen even more. Tropical soils are no more "peculiar" than other soils. But some agricultural people have not studied them much in comparison to those of the middle latitudes. Yet many of the tropical soils that have been studied give very high yields of plants that respond well with short days (Jurion and Henry, 1967 and 1969). Actually much more is known about tropical soils than many suppose but not all of it is in the English language.

Research for animal production is even more complicated. To the efficient production of grain and fodder, we have the added problems of insects and diseases that attack the animals. These must be controlled as well as those that attack plants. Then too, different kinds of animals require different feeding at the various stages of growth. For success, the proper breeds must be selected and adequate nutrition for the animals must be understood and the results taught. The problems and potentials are outlined by Ralph W. Phillips in the book edited by Moseman (1964) cited above. Thus for rewarding research in farming systems, many kinds of specialists who can work together are needed.

Some semantics of research

Many engaged in research talk about "basic" and "applied" research, yet few agree on definitions. At least one can assume that basic research seeks for the most nearly fundamental principles and quantitative explanations of phenomena that current techniques make possible, while in applied research these explanations are presumed to be correct and are applied to the solutions of problems that people have. Yet many have been led into basic research from the challenge provided by an applied problem that failed to yield to the scientific principles available at the time.

Somehow a few scientists feel that basic research gives the scientist engaged in it higher prestige than others. Research cannot be so easily catalogued in practice. Actually one should need to be able to look into the mind of the scientist to discover his intentions in order to make a clear distinction. Some of the most important principles were discovered "by accident" in trying to solve problems. Of course these "accidents" seem always to be observed by keen and resourceful minds. Unhappily some so-called basic research appears to qualify as such *only* because it has no application.

The imaginary "status" of basic research has been a serious handicap to establishing reasonable priorities for research efforts in some of the less developed countries. Some research scientists insist on working on problems emphasized at international meetings of scientists dominated by members from the advanced countries, even though the results have little or no application in their own countries where they live and work.

In fact one could give three categories of research (i) original basic research, (ii) original applied research, and (iii) nonbasic, nonapplied research. It is this third category that scientific research directors must guard against, even though those engaged in such exercises get their "papers" published (Kellogg and Knapp, 1966, pages 151 to 182). High competence and originality are required for success with either basic or applied research. Many research undertakings must include both, especially where presumed basic principles fail and new ones must be formulated to get on with the job.

Given imagination and industry, the most important qualification of the research scientist is his ability to diagnose problems, to state the critical questions clearly, and to design the critical experiment, whether within the laboratory or in the field. Actually both nature and the cultivators have already laid out many experiments; the scientist has only to find how the phenomena came about. For such knowledge to be useful, all relevant factors controlling the event must be taken into account. And the question is not primarily "why" but "how" and especially "how much."

Role of the basic research scientist in diagnosis

As agriculture changes, new combinations of practices are called for and new problems are certain to arise. As geneticists develop high-yielding varieties of cereals or other crops and these are introduced into other areas, what new diseases and insects may attack them? No one can predict. Not all diseases and insects are yet described in the books. As we have mentioned earlier, there is danger in adopting a new variety of a crop widely on many farms. A new pest can cause a near calamity by spreading rapidly from field to field. The new crops must be watched by first-class scholars in entomology and plant pathology. The same principle holds in economics and all the other specialties. Not all kinds of soil have been seen and studied.

This role of the research scholar is sometimes overlooked. And to maintain his scholarship he must be engaged in basic research. In most of the sciences related specifically to farming, especially to soil, crop, and livestock management, the basic research scientist who can diagnose problems must also be a traveller as well as a reader. He needs to see his subjects under different conditions. He needs to talk with his counterparts in other parts of the world. As an example, all plant breeders of farm crops anywhere in the world, whether they know it or not owe an enormous debt to the great Russian scholar, Academician N. I. Vavilov, who did so much to locate the areas where our various cultivated plants originated.

If too narrowly specialized, the scientist may tend to oversimplify problems in terms of his own narrow experience. What may be suggested as zinc deficiency or excess salts, for example, may turn out to be nematodes, or the other way around. For diagnosis, we need specialists with broad points of view.

Every country needs at least a core of these first-class scholars to avoid epidemics, horrendous errors in both soil and crop selection, and very great wastes of labor and funds in new land development. Once the problems or potential problems have been diagnosed, others may be able to go ahead with the adaptive research, planning, and advisory work.

Organization and purpose

Since areas needing expanded research for agricultural development vary widely in size and current development, no fixed pattern is possible.

Certainly it is helpful to have a central headquarters for a province or country of moderate size. Where practicable this may be associated with a strong university or other existing strong research facilities. The general rule is to build on existing strength in scholarship, library services, and other essential services for carrying on research. For example, where sophisticated electronic and other equipment must be installed and kept in repair the availability of local services is essential.

Good library services are especially important to make full use of the principles and data already available for similar kinds of soil, crops, pests, livestock, and so on. Actually the knowledge we acquire by reading is by far the cheapest we come by.

Principles of the responses of different kinds of soil to combinations of practices, including kinds of crops, diseases, insects, and weed control, water control, and the like can be transferred between countries, but *not* necessarily the specific practices or "models" of farming systems. Applications of the principles transferred must be related to the local skills, food habits, markets, and culture of the area. This requires some combination of adaptive research, field testing, and demonstration.

Some basic research may be required at the center in order to go ahead with adaptive research. The soil survey work should be headquartered at the center near the essential library and laboratories. Yet successful work in this field requires participation by other research centers, especially those at universities.

Outlying substations

Outlying substations are needed for carrying out some kinds of basic research as parts of wider programs, for dealing with any peculiarly local problems of crop and animal adaptation. Ideally these should be located to represent a general soil region. In practice, some compromises may be needed between soil association boundaries and provincial boundaries.

Especially for field experiments, it is important to select areas that are fair representatives of kinds of soil worth sampling. Every field plot is, in fact, whether realized or not, a sample of some kind of soil. Enormous funds and even years of skilled manpower have been wasted by ignoring this principle. I have seen elaborate field experiments laid out on insignificant kinds of soil. Even worse, I have seen them located on complexes of contrasting kinds of soil with soil boundaries within plots and between plots for comparisons of practices! No amount of "statistical" manipulation of the "data" can lead to valid conclusions about how any of the soils in the complex respond to management. Nor, of course, can useful experiments be laid out for comparisons of treatments on unlike kinds of soil, unless each plot in the set on any one kind of soil is representative of that one kind only.

Such enormously wasteful errors in research planning have resulted from failures to examine the soils in advance. Some were areas used for experiment only because the land was given by a donor or because it was "convenient" to existing roadways. These sorts of errors in location of experimental areas have made use of the "data" impossible because there is no way to classify the results by kinds of soil for guidance to advisory officers, cultivators, or engineers. Nor can people elsewhere use the "data" if the soil identification is simply "Ghana soils," "India soils," or "Argentine soils."

It has commonly been difficult to furnish library services to the outlying stations, especially, for the many important journals that need to be scanned. Recently, schemes have been developed for sending out copies of the tables of contents of journals to small stations. Copies of specific articles may be requested. With methods to use both sides of the sensitized paper a minimum of filing space is required. A system has been described for East African stations by S. Cooney (1968).

General versus special stations

I realize there are needs for small specialized stations to deal with certain crops, especially industrial and nonfood crops, such as tea that does well on certain kinds of soil not well-suited to most high-yielding food crops. I think, however, that too many have been established. Large well-financed stations can afford a full staff with geneticists, entomologists, plant pathologists, soil scientists, economists, engineers, and so on. Yet where funds are limited, the many interactions between soils, crops, insects, diseases, animals, weeds, water control, fertilizers, tillage, and the other factors tend to be seriously neglected. Also other better alternatives can be ignored.

The situation varies widely within the newly developed countries. The older dependencies of the United Kingdom, France, Belgium, and The Netherlands, for example, inherited some well-planned research stations. To varying degrees schools and colleges were organized for training local people as technicians and some went to Europe for advanced training.

As one example among many, early development in the Republic of Zaire (Belgian Congo) was carried on by King Leopold III of Belgium, mainly as a private effort under the "Congo Free State" (État Independant du Congo) beginning about 1885 following the discoveries of H. M. Stanley. Late in 1908 it became a colony of the Belgian Government. Before much could be done,

Belgian was overrun in World War I. Some agricultural research was initiated about 1926. King Albert took a great interest in the area and made personal study trips. About 1933 all the agricultural research was organized under the INEAC (l'Institut National pour l'Étude Agronomique du Congo Belge).[2]

This research was organized at a large well-equipped headquarters at Yangambi with leadership in each field of work for research in Ruanda and Urundi as well as in the Belgian Congo. In addition research was carried out in about 46 field centers and with other stations in many local areas for adaptability research. Remarkable progress was made in many lines of work to increase yields, to develop new adapted plants and animals, and improved systems of management. Fortunately the results are published in various series in Brussels and were recently summarized by Jurion and Henry (1967, 1969). Since these published results represent many kinds of tropical soils and climates they are valuable for guidance in the other tropical countries.

It would be hard to suggest a more urgent task in agricultural research for Africa than to carry on the program of this great institute in the Republic of Zaire. During the earlier development many local people were trained in the work. Similarly many were established by the French and British in Africa. In fact, agricultural research has been strengthened in many of the less developed countries. This improves their opportunities a great deal.

India has several useful experimental stations and research centers that owed their origin to the early British efforts as well as new ones. Many of these have been expanded and strengthened, and new ones added by special efforts of foundations and help from other governments.

2. The word *agronomy* has several unlike meanings as used in the United States. Most commonly in American literature it refers to both basic and applied scientific work with field crops. Some use it for only the applied work and the basic research is included under plant genetics, pathology, or physiology. At the other extreme, some also include under agronomy soil management and even basic soil science intended for application in forestry, horticulture, engineering, and town-and-country planning, as well as to field crops and pastures. The equivalent word in French usually refers to *all* basic and applied research in agriculture, forestry, and perhaps fisheries, including the full range of social and natural sciences dealt with by an American college of agriculture.

Combined roles of research and education

All universities have the basic responsibilities of storing the accumulation of human knowledge, of teaching it, and of adding to it. This does not imply, by any means, that all research can or should be done at a university, but most professors need to have some hand in it to keep themselves and their teaching up to date. In planning research in agriculture it is important that professors and their students have opportunities to see how some of the ongoing research is carried out. Where practicable, some research financed by provincial or national governments should be located near universities. There are advantages to the research scientists, to the professors, and to the students.

Many of the most distinguished professors and scientists in the world have combined responsibilities to do both teaching and research at the same time. Actually arrangements in the American land grant universities for scientists to do both research and teaching have been beneficial to American agriculture.

A high proportion of the leading agricultural scientists in the developed countries have grown up in this atmosphere as students, graduate students, and research scientists. There has been considerable interchange between universities and government. In fact much of the research of the U.S. Department of Agriculture is located on or near univeristy campuses. Many former professors have worked with the U.S. Department of Agriculture, the Tennessee Valley Authority, and other governmental agencies. Also many interchanges of scientists take place between nongovernmental agenices in agriculture, including many of the great foundations now offering assistance abroad, and both the U.S. Department of Agriculture and the universities.

All this helps to develop specialists in research with broad points of view. A teaching assignment forces the scientist to organize what he knows and, of course, what he does not know. As a personal example, most of my early career was in research. Then I had to organize courses and teach students. Nothing could have stimulated my reading more. My own research had covered only a fraction of what had to be included in the courses I gave. Others have told me of similar experiences. Then too, research experience leads to emphasis on the quantitative, not simply general qualitative principles.

Universities are not well equipped to give students training except in those fields where they have excellent professors and going research programs. Nor can all of the students' study in most fields of agriculture be limited to the university campus. Graduate theses based solely on laboratory or other inside study, without clear ideas of the basic purpose of the research undertaking and its application are not always helpful.

Some have said that research takes such concentrated effort, that many scientists cannot do both teaching and research. No doubt this is true for some. And no doubt some teachers lack the patience for the deep study required in research. Not all can do both. But those who cannot should certainly not be selected to undertake technical assistance for helping people in other countries to establish good research programs.

Advanced education abroad

Many scholars have benefited from education in other countries. During the Middle Ages when Latin was the *lingua franca* of European scholars there was considerable movement of both students and professors (Rashdall, 1936). No doubt this system did much to promote common cultural attitudes and values in Western Europe which were inherited by European settlements in North America and elsewhere. And such beneficial exchanges have continued, especially for advanced study. In more recent years many came to Europe from Asia, Africa, and elsewhere.

With the emphasis on technical assistance to the less developed countries, great numbers from the tropical and subtropical countries have come to Europe, the United States, and Canada for training in all aspects of agriculture. Some came for B.S. or M.S. degrees, and many stayed on for the Ph.D.

Fellowships, and scholarships, or grants have been available within some of the less developed countries. Then too, people of private means in these countries have supported a good many students abroad. Others have had substantial assistance from the host country by national agencies carrying on technical assistance abroad, by universities, and by private foundations. The United Nations agencies have also given support to many such students. There can be no doubt that much of this has been useful. Yet there is reason to think that some has been a great deal less than that.

First of all, many such students were not prepared in the language of the university to understand the lectures clearly, to write clearly, or to read rapidly.

Secondly, they were moved into quite a different social pattern with little time for adjustment. In some developed countries the dormitories seemed most luxurious. Many had stipends sufficient to buy a car. Unconsciously, perhaps, their associates helped to make "Americans" or "Europeans" out of them. With 3 to 8 years of this environment, they did not want to go home. They looked for opportunities in North America or in Europe, or within American or European companies. Many found opportunities elsewhere.

Such shifts have been a burden on the less developed countries. The students had had the opportunities for education through high school or university at home, at considerable local costs. The losses to some of the less developed countries from this kind of "brain drain" have been large. So as time went on many governments of the less developed countries and of host countries insisted that they return. At first, the returning student might not have a job, but with the new rules about return many governments were committing themselves to a job for him.

Although the student had a degree in an agricultural science, much of it dealt with different soils, crops, animals, pests, machines, marketing, food habits, tax methods, trade, governmental organization, and other items than those in his own country.[3]

Not only that, the feeling of prestige from having an advanced degree from a well-known American or European University may make communication with local cultivators at home difficult, to say nothing of dressing in outdoor clothing to show cultivators how to do a job better or to study a problem in the field.

These statements do not apply to all. In the less developed countries I have travelled in the field with scholars who had their advanced degrees from Europe, the United States, or other developed countries and who could communicate well with their own farm people. I hasten to add that some Americans and Europeans

---

3. Some of these difficulties as applied especially to economics are very well stated in a book on the *Teaching of Development Economics* by several scholars and edited by Martin and Knapp (1967). I recommend it highly, and especially the first chapter, *The limitations of the special case.*

who go to a less developed country cannot communicate well with cultivators either.

Let us hope that a higher proportion of the funds now used to send such students abroad will be used to help build universities in the less developed countries themselves, including assistance for graduate schools, such as the Rockefeller Foundation and other foundations have been doing in Southern Asia, South America, and elsewhere.

Most training of professional agricultural scientists can be done best locally. Students can live and study in the social and biological environments to which they are reasonably accustomed and in which they are likely to work. Demonstrations and exercises in problem diagnosis, which is so important, can be carried out nearby. Interactions among local soils, crops, livestock, and both economic and social conditions can be explained more clearly.

These graduates can then be tested in local positions. Those demonstrating technical competence in problem diagnosis in the field, resourcefulness, and ability to work with cultivators, can then be selected for graduate study in their own country. These graduates can be appraised again after work in the field at home. They are more mature. Most will be married. The good ones will have jobs.

From among these selections can be made for further study abroad to gain special skills. This does not necessarily mean at a university or on a quest for a doctoral degree. (Scientists should be judged by their competence, skill, and productivity, not by their degrees.) Many of these tours could include study in industry, cooperatives, universities, or governmental agencies with successful programs, in the developed or the less developed countries.

Eventually every country must train most of its own scientists, engineers, and technicians. To develop their resources the need is great and will continue to be. If countries are to be truly independent they will need to depend mainly on their own citizens for scientific and technical advice.

Thus I should recommend that the development of first-class colleges and universities, with associated research, should have very high priority for the use of scholars and funds under technical assistance programs for economic development. This does not mean that exchanges of scientists and scholars should be discontinued, not at all. Such exchanges among countries can be beneficial to all, but we should aim to have exchanges on even terms.

International research institutes

Excellent reasons suggest the economy of research institutes for common problems to be developed by groups of small countries. Yet these are difficult to sustain by funds collected from sovereign governments. Since the institute is bound to be located in some one of the countries it is difficult to convince a government that it should make appropriations to a facility in another country.[4] Some have been supported in this way for a while, with part of the support coming from a developed country or a UN agency. For example, the International Atomic Energy Agency of the UN, located in Vienna carries on some basic research on the uses of tracers applicable in studies of the uptake of nutrients from air and soil by plants. For several years the United Kingdom supported an important research facility useful to several countries in British East Africa—the East African Agricultural and Forestry Research Organization at Nairobi in Kenya.

Many of the most important new research institutes supported by the developed countries for the less developed areas were those depending mainly for funding on experienced private foundations, now commonly with additional help from governments sponsoring foreign aid.

Henry A. Wallace, Secretary of Agriculture of the United States during 1933 to 1940, and Vice-President following that, was deeply interested in Mexican agricultural development and in maize breeding. The yields of maize, an important food crop, were low. He pointed out the enormous potential to the Mexican government and to the Rockefeller Foundation. His enthusiasm was acted on by both (Stakman et al., 1967). The great success of this effort is well known, not only in breeding maize and wheat for higher yields with better water control and fertilizer, but also for taking account of soil management and disease control.

Besides the institute in Mexico (CIMMYT) an important one has been established for tropical agriculture in Colombia, S.A. Jointly the Rockefeller Foundation and the Ford Foundation initiated and now support the International Rice Research Institute in the Philippines and the newer International Institute for Tropical

---

4. In fact, it is difficult for states in the United States to pool funds for a local station on soils common to both. Neither state legislature wants to send appropriated money outside its state. Hence most regional experimental centers must be financed mostly by the national government.

Agriculture in Nigeria. Both these and other foundations have assisted many other countries to develop agricultural research stations.

These efforts by several private foundations have made it possible to have well-staffed research centers that help to solve problems in countries, besides those where the institute is located. Many of the problems of bilateral assistance by governments or even of multilateral efforts by FAO have been avoided. Also U.S.A.I.D., Canada, and agencies of other governments have helped in the important efforts to develop research programs in the less developed countries.

So far, however, many of the efforts on food production have emphasized the grain crops. In some of the less developed countries the people have been long accustomed to other foods. These too need improvement, and recently progress with other crops has expanded. Also the opportunities for many cultivators to rise above the subsistence level demand skill and guidance in the production of industrial crops, such as cotton, rubber, African oil palm, and many others that earn them cash and give foreign exchange to their country.

Requirement for trained advisory (extension) officers

The training of senior advisory staff in agriculture is roughly similar to that for research scientists, although a bit broader and more in communication skills. It is very helpful to both the advisory specialists and the research scientists to be housed together or in close proximity. At least they should have rather frequent meetings.

The advisory officers can bring problems needing attention to the research staff and, of course, they must be counted on to give the information gained by research to the cultivators, directly or through the local advisors. As mentioned earlier many advisors can also take major responsibility in the field for adaptability research and some of the field testing (Kellogg and Knapp, 1966).

Advisory officers also have a role to play in helping cultivators to form groups or associations for self-development through group meetings and visits to field demonstrations that have application to their own management systems.

7

Communications

In this chapter the aim is to present broadly some of the problems basic to communication between people of unlike cultural backgrounds. People in the world differ in language, education, occupations, religion, household arts, and in many other ways. The potentials of the soils they have, the local climate, distances from neighboring people, availability of minerals, their aptitude for trade, and so on have a lot to do with the skills they develop in order to live.

These and other cultural differences make the world more interesting, but they can lead to serious misunderstanding through failures of communications.

Since World War II, we have had many more "nations," including many from Africa, join the UN. If by *nation* we mean a group of people with reasonably common customs, language, and motivation for freedom, this is hardly true in some except in a legal sense (Andreski, 1968).

Among the peoples who were in North America when western Europeans arrived, there were large differences. There were large differences, of course, among immigrant groups who came later. Except for those taken against their will from Central Africa, most who came to the United States had reasonably common heritages from Europe. Yet these too, were not identical in language, customs, or common law.

As a result, there are continuing problems of communication. Even within one language, words do not have the same significance to all speakers and listeners. People do not necessarily have the same interest in unlike ideas or goals, nor do they attach the same significance to them.   This is true at all levels of education and social strata.

The situation among educated people in Western culture has been especially well stated in Lord Snow's *Two Cultures* (1959). He was by no means the first to explain the large communications gap between present-day scholars in the sciences and in the humanities.   Then too, we have serious gaps between many scientists and engineers, between social scientists and natural scientists, and between poets and cultural anthropologists.

During the early nineteenth century, most university students in Europe and North America took similar courses of study.   They studied the classics, theology, mathematics, logic, history (including the history of art), literature, and philosophy; and especially near the end of the century they studied natural philosophy, or what we now call natural science.   Various of them became army officers, scientists, physicians, poets, public officials, and so on. Generally, graduates were able to communicate with one another without so many of our present-day difficulties.

But the classical curricula of the eighteenth and nineteenth century-universities became rigid.   Reactions set in, especially in Germany and the United States.   Germany placed early emphasis on specialization in science and engineering and we in the United States made strong efforts toward technology.   Our land-grant colleges and universities were established in the middle of the nineteenth century to make higher education available to the sons and daughters of farmers and workmen.   Emphasis was placed on the specialized application of the sciences to engineering, agriculture, and the household arts.   Today many of these colleges now give great emphasis to basic science and to postgraduate education.

In recent times we find people with different kinds of advanced education and many with high specialization in scores of different narrowly defined fields.   Efforts to build a broader base of education either in advance of specialization or along with it have been attempted. We can hope that these trends may be accelerated. Yet looking over our educational history, it appears that we are far more able at increasing specialization than we are at broadening education.   This seems to be equally true in both the humanities and the sciences.   This may be partly because specialization is easier.   Broadening education at the expense of depth is not the answer.   Present-day students do need the discipline of deep scholarship and specialization as well as breadth of knowledge. There is nothing about specialization that makes people narrow-minded, but a great many narrow-minded people tend to specialize too narrowly. If they are permitted to remain narrow-minded, the fault lies with their professors and the educational system.

Actually today, except for a few near the very top in scholarship and broadly educated, many men and women graduating in the sciences or in the humanities find communication with one another difficult.

First of all, some knowledge of higher mathematics is essential to an understanding of basic physics, chemistry, and biology. Fewer people graduating in the humanities today have had education in the calculus than was true at an earlier time. Those who do not know what the symbols of mathematics mean, and without practice in their use, have extreme difficulty in understanding modern science. On the other hand we find a similar lack of appreciation of poetry and other literary arts among many of our scientists and engineers.

In other words, to read or to listen, one must understand the symbols of the language. In their work, scientists are most familiar with one set of symbols, those in the humanities with another set. Fortunately we do have many people between the poles who are familiar with both sets. For those unfamiliar with science there is little possibility of understanding the industrial revolution, let alone appreciating it and accepting it. Those unfamiliar with the symbols of the humanities are shut out from understanding a great body of knowledge, accumulated over the centuries, that bears on art, morals, justice, and agricultural development.

Many of the literary intellectuals recognize the bad things in the world but seem to assume that they are mainly a part of the inevitable social situation in which man finds himself. Thus, they fall into a kind of moral trap. They see man as a lone being without recourse; each person has his unique tragedy. And this view is not entirely wrong.

On the other hand many scientists and engineers overlook the fact that an educated man needs to understand three sets of relations: (i) the relations among things—the field of science; (ii) the relations between man and things—the field of art, including the industrial arts; and (iii) the relations of man to man—the fields of justice and morals. The great literary artists, poets and novelists, as well as the theologians, historians, and other scholars of the humanities, have given very important insights into these relations. Each educated man and woman should have a reasonably balanced understanding in all three areas.

Yet the edifice of understanding of the natural world that science struggles to erect, in its intellectual depth and complexity, is one of the most beautiful creations of Western culture. To ask a natural scientist if he could say something about the second law of thermodynamics or Robert Boyle would be like asking a poet if he had ever read anything by Shakespeare or Joyce, or asking a social scientist if he is familiar with Edmund Burke, Vico, or Pareto. The educated man between these two poles in our culture would, of course, be familiar with many of the classics of natural and social science and of the humanities.

To turn briefly to the "industrial revolution," it is hard to put a date on its beginning. What we call Western science, which owes much to Greek, Roman, and Arabian thinkers, began at about the same time as the great geographic discoveries—about the time Marco Polo made his famous journey to China. The great discoveries of Leif Ericson, Columbus, and many others stimulated both science and invention. Although modern European invention and science started at about the same time, they did not merge until later. Most of the inventions in agriculture and industry were made by skilled farmers and skilled craftsmen. With several notable exceptions before the middle of the 19th century, relatively few of our inventors were scientists, as compared to now.

Before science had proved its practical value, many people in high positions fortunately became deeply interested in it. Charles II of England chartered the Royal Society and that Society the Royal Institution. Thomas Jefferson and others thought that men should be trained in science. And so did the leaders of Germany in the early nineteenth century. Then in the twentieth century science and industry began to merge, especially after World War I. As industry became more complicated, invention by the bright artisan or brilliant scientist became more difficult. Now most industrial and agricultural research is carried on by teams of scientists and engineers. But along with this merging of science and invention we began to have some schisms within science itself.

Different cultures

Most of us who grew up in Western culture after the industrial revolution, have generally accepted certain religious, economic, and political views. Most take for granted our own mores and values, even the silly ones, without critical thought. Obviously the problems of communication with people of another culture require considerable study and patience.

People of other cultural systems have different religious views. Few people of the Arabian culture, for example, can conceive of secular citizenship, which we take for granted. Similar differences in attitudes account for believing that the Chinese, the Vietnamese, the Indians, and others may not be doing what we regard as sensible. Actually these other people have some values superior to ours. The educated man tries to understand their values. A few are against modern technology. For these countries to develop their agriculture and industry to be as productive as those of the Western nations, some of their ideas must change. Yet it is equally important that we recognize those values that do not need to be changed. In working with them we should suggest as few changes as possible. They do not want to become some other kinds of people. They want to maintain their own integrity. Also we will be better off if we recognize which of their ideas that are better than ours. So here I am not arguing for that easy virtue of tolerance but for that far superior virtue, appreciation.

Problems of communication are even greater in the most primitive parts of Central Africa and elsewhere. A large part of what people in advanced cultures have learned has come to them through their eyes from printed newspapers, magazines, and books along with illustrations in perspective. This too we take for granted. Among primitive people much of their knowledge comes to them through their eyes and ears. Yet many cannot read or even understand a picture in perspective, on a poster, or in a movie (McLuhan, 1962).

Many of these people have been isolated until recently from the main stream of world thought, art, and industry. They are illiterate and highly religious in the sense that little happens by chance. Everything is ordained by good or evil spirits. Some of their views are highly traditional. Their societies tend to be closed. Some are suspicious of strangers. Many of their visitors had come to steal, to raise their taxes, to get their boys for the army, or to take their land. If someone visits the village, they cannot help wondering about his intentions.

The people in these societies think differently from the way people brought up in Indian, Chinese, Arabian, or Western cultures think. We can fail to help them if we assume that their thought processes work like ours do. They have their own words for kinds of soil (Levy-Bruhl, 1966). Many are more perceptive of local plant differences, for example, than most of us. Most of their local languages are mainly oral and reflect, like all languages, the things most important within *their* culture.

We can help them and work with them with patience and advanced study. One never begins with criticism. Certainly any imaginative person going to such a village for a talk with the chief, can find something on his way through the village to congratulate the chief about.

The gap between the rich and the poor may be enormous. Differences are very striking between countries where most people are poor and countries where most people are reasonable comfortable. And of course this gap is understood more by the poor than by the comfortable. In a way the poor countries are in revolution against the West; but it is a curious revolution since they want to develop many of the skills we have.

Most of us have accepted the idea that the poor countries can and should increase their efficiency in agriculture and in engineering. Yet some of us are only beginning to realize that these people do not all want the things that we may assume are good for them.

Development for the people of many areas can come about only with some change within themselves. They need to be dedicated to their own goals and to work. During the past in many African countries, the real authority was distant, in Europe, or now, far away in a port city. Somehow these distances must be overcome for communication to be effective.

Many of the economic principles taught in Europe and North America are unconsciously biased by our culture (Martin and Knapp, 1967). For example, among very poor cultivators a rise in prices for food crops may result in less being sold, not more. They sell what they must to get a minimum of cash. If they can get the same money for fewer bags of rice, for example, some hold more at home to eat themselves. This kind of statement is commonly belittled by the poorly travelled Western economist because it suggests a "backward sloping supply curve"! This is contrary to his theory. Yet it happens.

Many have jumped to the conclusion that the poor cultivators they see for the first time are "lazy." When there is no urgent field work to do, one sees them resting in the village. The tourist assumes they are lazy. But the tourist probably gets 3 to 4,000 calories a day, whereas the poor cultivator may not have more than one-half that. If anyone of us were that short of food and had spare time, we would be unlikely to use it for golf, tennis, or jogging!

Clans and politics

Andreski (1968) explains the concepts in Africa of "nations," "ethnies," "tribes," and "clans." The peasant owes obligations to his clan, and they to him. When many of these countries received their independence, some people lost face by recognizing a person from another ethny or tribe as head of state compared to a former governor from Europe. In fact, some belonging to other ethnies or tribes somehow felt left out when independence came! They assumed, perhaps with reason, that appointments within government would go to members of other tribes and clans than theirs.

As the young men and women in such countries had learned to read and even to get higher education, most had little thought of returning to their village, but went to the capital city for a new position. If someone from his clan was in a good position there, he had a good chance, but not if he lacked the traditional entry into the bureaucracy.

These sudden changes have caused difficulties in communication by making ineffective the older traditional loyalties of society without yet establishing new ones. In the meantime accomodations that appear to Westerners as "graft" or "corruption" arise. (And some Westerners contribute to the graft!) Some considerable patience is required to help the people establish new relationships that lead to effective communications and programs for development that benefit all. Strictly aping of Western nations is unlikely to be a good longtime solution.

Town-and-country relationships

We have already mentioned that for our current time it was unfortunate that in Europe, the United States, and elsewhere many people were concentrated in the port cities. Possibly this was nearly unavoidable so long as heavy transport was by ship. Then with the industrial revolution, most manufacturing was also concentrated in the same cities, including the industrial sectors of agriculture. Much of the processing of farm products and the manufacture of modern inputs for farming took place in the big cities. As farming improved, gradually farm labor moved to the city with much of it going into the industrial sectors of agriculture. Now fortunately many are aware of the advantages of smaller towns where farmers and other workers know one another.

Transport by railways, streams, and highways is essential to develop farming, mining, and forestry. The industrial sectors of agriculture can be located in the country at primary and secondary centers. The inputs for farming can be manufactured at such centers and also the processing can be done there. Country towns can serve these functions, much as the early villages did, and which many do now.

I have visited many rural villages or small towns in Asia where cultivators live along with wagon makers, leather workers, metal workers, weavers, and several others. The essential schools, hospitals, stores, and so on can serve the people in the town and in the rural trade area as well. We shall return to the importance of rural towns again for other reasons. Such centers can aid enormously with communications among people of different occupations and reduce the great distances between the cultivators and the places where decisions are made or could be made.

Between cultures

One can transfer the results of research and experience on named kinds of soil between countries and continents in order to estimate potentials for use. But farming systems are developed and used by people, and what they can and should do also depends on their social habits and goals.

To improve existing farming systems or to develop new ones requires great skill in science, technology, and communications. When the United States undertook a program of technical assistance under Point Four, many said that skilled American farmers and county agents should be sent abroad to show other people how to farm as it was done in the United States! Probably some could do this, but not many had the skills of communication to do so in countries with soils unlike those they were accustomed to and in cultural systems of unlike goals, status symbols, and all the rest.

Most of those successful at such technical assistance must be good "operating anthropologists" who like people and enjoy seeing them develop suitable systems in their own way. Of course, some old traditions do need to be changed for modern science and technology to be effective. With tact and demonstration progress can be made if only the really few highly significant changes are suggested at the start. And in some places we may need to begin with only one change.

Some people from the developed countries, unadapted to such advisory work, have been sent on missions to one of the less developed countries. On their return they have said to me— "nothing can be done without a complete change in the whole organization and culture of the country." One does this only with a revolution, with the outcome very doubtful.

Among many people in India, animals are held in respect. Certainly this is no more irrational than many Western views. If a person helping them with their farming systems cannot accept such a concept and insists on hunting animals with a rifle, he is clearly unadapted to the work he has undertaken. Such a person must also be careful of how he uses his camera. Some people very much dislike having their "pictures taken." And so we might go through a long list of customs that the outsider needs to learn as much in advance as possible, if he is to be communicative, sympathetic, and helpful.

In some countries, both developed and less developed, differences in religion or race handicap cooperation and development for parts of the country. Many of the difficulties may have been inherited from very much earlier colonial rule. R. E. Baldwin (1966) gives some examples from a study of economic development in former Northern Rhodesia that apply in other states in the southern part of Africa, and elsewhere. But examples may be found even within some of the "developed" countries. If the good jobs, education, and other opportunities are limited to certain immigrants or certain classes, the other people may have little hope.

The planning of development in an area or country by and for the people, necessarily becomes a part of the political process. Some priorities must be established. A poorly made irrigation plan, runoff control scheme, or land reform project makes people lose confidence in the "new technology" because it is bound to fail. Attitudes on land tenure are old and not easily changed quickly. Thus all the way along in helping people with agricultural development, communications are vital. Nor can one accept fully any one of the many rigid interpretations for specific "stages" in the process. Gunnar Myrdal has many useful discussions on aspects of this process in his *Asian Drama* (1968).

8

General services

A few general problems and services are briefly outlined in this small chapter only so they may not be overlooked as vital factors in the agricultural development of an undeveloped area.

In one of his lectures of about 1755, Adam Smith wrote

Little else is requisite to carry a state to the highest degree of opulence from the lowest barbarism, but peace, easy taxes, and a tolerable administration of justice; all the rest being brought about by the natural course of things. All governments which thwart this natural course or which endeavor to arrest the progress of society at a particular point are unnatural, and to support themselves are obliged to be oppressive and tyrannical.

In terms of present day problems of helping to develop modern farming and the infrastructure for modern agricultural systems, this may seem to be a bit oversimplified but still it contains a great deal of truth for today.

Legal services

Existing common (or tribal) law and statute law on land and water in some countries does need amendment for proper use without possible bad effects. Uncontrolled burning of savanna, cutting of forests without replanting on soils unsuited to other uses, pollution of streams, building roads, houses, and other structures, on unstable soils or building so densely as to initiate floods, are a few examples of matters needing attention in many countries. Unfairly low prices to cultivators by private or public buying agencies, including export taxes on agricultural products, reduce agricultural development.

Taxes

The burden of general sales taxes and general import taxes bear most heavily on the poorer people in both the developed and the less developed countries. But in the ldc's such taxes on luxury goods fall on those most able to pay.

What investments count

Cultivators must be able to accumulate capital for investments in fertilizers,[1] machines, and other inputs. For any one cultivator the amount invested may not seem like a great deal. But suppose most of them bought two or three more bags of fertilizer, a pesticide sprayer, and a small garden-type tractor, the total effect would result in a great boost to the economy, compared to a beautiful opera house or even a well-placed hydroelectric dam, perhaps. Dams, large buildings, highways, and so on are highly visible, many are very helpful. But the total potential investments of progressive cultivators, which may be considerably greater, are commonly ignored. Another great need is a good storage facility in each village or town that handles farm products.

These kinds of investments are not commonly considered, yet the "multiplier effect," in the sense of Lord Keynes (Stewart, 1967), of these investments in farming may be large for the economic growth of the area or the country.

Statistics

Statistics on farming, food consumption by cultivators, farm employment, and especially employment in the agricultural industries and services, are poor for the world and are especially inadequate in the less developed countries. In the developed countries several times as many people work in the agricultural industries and services as work on farms. As the less developed countries move ahead, farm workers will also move off the farms into the other sectors of agriculture.

---

1. By some curious quirk of reasoning that may have seemed valid 100 years ago, most statistical reports still refer to "fertilizer consumption." Actually people *use* fertilizer, but as an investment to improve the soil and crops. If properly used both the direct and indirect effects on the soil are valuable in the future as well as in the year applied.

In making estimates of production, altogether too much emphasis may be placed on those few large items that enter international trade, such as cereal grains and industrial crops. The published world figures on *food* production are certainly lower than what people as a whole actually eat. In many thousands of small markets no one has an accounting of trade. Any planning based on most such current figures is unlikely to come out well.

Health services

Health services are also inadequate in many areas. Malaria has been perhaps one of the biggest farm problems in warm countries. People with this disease are poor workers and usually die young, after only a short working life, because of the high percentage of their lives spent in helpless infancy and "old" age. Fortunately, DDT has greatly reduced this disease in many warm countries. Poor health can be a great drag on the economy. One of the great opportunities for improved health is through the introduction of a wider variety of food crops and livestock. In a few areas one must move cautiously with greatly increased protein without good medical advice, since people not having had much protein in their diets for generations may not be able to digest so much as is customary in the advanced countries. Many other examples could be cited.

Savings and social security

Savings are also difficult or impossible for many cultivators and other rural workers in the less developed countries. There are few reliable country banks and little or no protection against theft. This is illustrated dramatically near the margins of the hot deserts. During the high heat of midday many are wearing very heavy long coats. But there are few if any places they can leave them without high risk of theft, and at night they need them. I have seen some poor men put their savings in gold and silver bracelets fastened securely around the ankles and wrists of their wives.

Thus without any reasonable place to save or to invest small sums, for many millions of cultivators the only social security is to have many sons. Because of the lack of public health and medical facilities, each couple needs several sons to be reasonably certain to have someone to care for them in their old age.

Population control

Birth control is difficult in many areas where people are poor. Until they have some form of social security besides many sons, and opportunities for saving, most are not likely to be greatly interested in birth control, even if facilities were available. Also in many areas daughters may be valuable since the "bride price" goes to the father, or to the mother in some matriarchal societies.

Because of the traditional need for many children the social "rules" favor large families. In some societies unless the wife has a live child within a year after marriage, her husband can get his bride price back. Then too, in many villages there is little or no privacy. Many have no "sanitary facilities," good or bad (Figure 82). The whole matter is entirely different for Western families with good incomes and modern housing. Even in the Midwest of the United States an old slogan was still current in rural areas before World War I "marry young and you will have a good home when you are old."

FIGURE 82

One can hardly say that sanitary facilities are good or bad in this village in India. scarcely exist.

FIGURE 83

As population increased in Assam, some hilly areas eroded down to old laterite, with coal beneath. A peasant women carries it up from the mine to the road.

These old mores change slowly of course. In several primitive underdeveloped areas where medicine was introduced with few efforts to improve agriculture or other sources of income, both population and poverty increased (Figure 83). Any hope of success in most of the ldc's for "zero population" growth, without war, will depend on better incomes, medical services, education, and opportunities for saving for old age in some form that is secure to the savers.

We must hasten to add that compulsory savings, especially for investments elsewhere, can work to the disadvantage of the poor cultivators. For them unfair taxes can also make them worse off and reduce their incentives for production and improvement. On these interrelated problems of savings and taxation in the less developed rural areas, P. T. Bauer (1965) has many useful insights as well as on other complex problems of economic development in the ldc's. He is generally against compulsory savings that may cause hardships to the cultivators who are neither seen nor even known by those who promulgate the schemes. Much harm has been done to cultivators by well-meaning bureaucrats who have never seen their problems nor understood their circumstances. In the less developed countries especially, one must go to sample villages and farms to have a reasonable basis for decisions.

Excessive birth rates certainly can be a barrier to economic development and to the improvement of services in areas of poverty. Yet without effective efforts toward development with good jobs and social security, perhaps not much progress on population control can be expected. Then too, a great deal depends on how the rewards of the hard work for development are distributed. Certainly the average per capita gross national product (GNP) is not a good guide, because the total of goods and services added into GNP, include many that are not helpful, leave out some critical elements of a good society, and may be distributed very unevenly among the population.

Price incentives

Prices are an important element in the development of farming. In countries with many poor people in the cities, policies are commonly proposed to keep food prices so low that cultivators have few incentives to invest in fertilizers, pesticides, and water-control practices for the higher yields that are essential for good supplies of food. On the other hand, of course, too high prices for farm products can reduce sales and even lead to surpluses in the market, especially where labor is poorly paid.

There has been a serious dichotomy in parts of southern Asia where both research and demonstration have shown clearly that production can be far higher than formerly with moderate investments in farming.

Similarly, improvement of farming has been slowed down by food imports from other countries at concessionary prices. On strictly humanitarian grounds such gifts for hungry people are praiseworthy. Yet they can cause delays in the permanent solution of the food problem within a country by discouraging the small but essential investments by cultivators to make substantial increases in their own production.

Infrastructure

With good prospects for production, storage facilities must be available in the local markets—villages and towns—where cultivators can sell their products. The local storage must be near enough to

accomodate local transport: carrying on the head, in a bullock
cart, in a canoe, by camel train, or some other principal way.

Storage problems are especially serious in warm countries that
are either continually moist or that have seasonal wet periods. The
Japanese have excellent methods for building such storage space
within the villages for rice and other grains. Ultimately, I look forward to the use of electric power for refrigeration that can be as
helpful for grains and other crops as for meat and fruits. In many
moist tropical areas the potentials for hydroelectric power are
enormous.

Besides these, in many countries with great seasonal variations, especially in rainfall, large storage facilities are needed in or
near the small and large cities for substantial carry-over supplies,
especially of grains and other food products, to be used in those
seasons that are unfavorable for good harvests.

Both local roads and either railways or truck routes for long
shipment are needed in many places. With adequate transport, advantages can be taken of the great unused potentials for farming
in the interiors of Africa, South America, and parts of Asia.

Minimum wage

Minimum wage regulations are needed in many countries to
avoid exploitation of labor in both city and country areas. Where
laborers must live near their work, away from their homes, the
regulations need to include sleeping quarters, food, blankets, and
sanitation as well as wages.

Now, of course, this does not mean very high wages for unskilled labor, especially because high labor costs could hold back
promising developments or lead to the jobs being taken by workers
from outside the area. Examples may be cited of development
held back by both very low wages and by wages too high, in both
the developed and the less developed countries. Many factors need
to be weighed to arrive at a just wage.

Police and security

Police and related security measures are needed in many areas
if development can be expected. Farm harvests, tools, and personal belongings need to be safe or incentives for production are lost.
This need is important along the whole route—in the fields and
homes, storage areas, transport, processing plants, and stores.

In countries where many people are desperately poor, security of goods is difficult. With a large percentage of unemployed without prospects of jobs or food, relief measures must also be taken. Although those problems are especially acute in large port cities receiving immigration of unskilled people from the villages, it can also reach far back into the country areas.

The problem lessens as people improve their lot and gain confidence in their local, state, and national governments. It worsens as people feel their government is indifferent to them and if many bureaucrats are themselves cheating.

9

Some ways toward agricultural development

Although many so-called developed countries have great opportunities for more rewarding agriculture, the great present need is in the so-called young countries. If their expectations are to be met within the next 5 to 10 decades, all countries—both developed and less developed—must continue to increase their ability to cooperate. There is no magic, no panacea. The closest thing to a "key" is implied by the word cooperation between countries and between urban and rural people within countries. In some countries racial or religious differences are still serious barriers to development. Countries must depend on one another. None has all the essential resources needed for its own development.

We hasten to add that cooperation does not mean uniformity. Few countries can be expected to escape their historical and cultural backgrounds, and those that do so completely could be worse off.

Even the several advanced countries have different ways of achieving cooperation and development in agriculture and in other sectors. Some have farming and processing in one collective unit. J. Tepicht gives an excellent statement for his country in his *Problems of the Restructuring of Agriculture in the Light of the Polish Experience* (Tepicht, 1969). And certainly Poland has a rich background of history.

All countries have inherited certain patterns of farming and of agricultural services and infrastructure. In Africa a part of the new countries inherited bad boundaries, many of which were

drawn in Europe with little regard either to ethnic or tribal boundaries, or to size. Some are almost too small for a viable domestic
market. Broekmeijer (1966) gives an excellent summary of the
characteristics of the developing countries and contrasts among
them.

Full development will take much new research and especially
the much wider application of already well-proved scientific and
technical methods. Both private and public investment and trade
among countries must be increased. The developed countries
must furnish a considerable part of the capital and some first-class
people to give technical assistance for helping to initiate the essential surveys of resources, universities, research, demonstrations,
and the like.

As we have already emphasized, education is essential for
agricultural development, not simply in farming but in all technical
and administrative skills necessary to operate the basic infrastructure for efficient agricultural systems. And the people need jobs.
As farming improves anywhere the need for on-farm workers de-
clines and that for manufacture of inputs, processing, transport,
and other services increases. The reduction of the enormous wastes
between fields and users of farm products is fully as important as
increased yields on farms.

Much has been written on the economics of peasant farming.
Both Manning Nash (1966) and David H. Penny (1966) have interesting statements partly because they have "been there." Also my
friend T. W. Schultz has some helpful insights in his *Transforming
Traditional Agriculture* (1964). Economics is very important in all
aspects of development, but ideas on the subject have changed a
great deal. I recommend Professor Joan Robinson's *Economic
Heresies: Some Old Fashioned Questions in Economic Theory*,
(1971). Others have written that agricultural development in most
of Africa and South America is about hopeless for many reasons,
including ineffective governments, poor educational programs, and
graft. Perhaps some of the pessimism is due in part to terms like
"development decade" when, in fact, nearly a century may be required. Points of view on Africa naturally vary widely. Few
writers on "Africa" have studied or travelled in all the countries.
Africa is a very large area. It is easy to make at least some serious
errors in generalization. The student will do best by using qualified
appraisals by individual countries.

Town-and-country development

As we have emphasized earlier, many still think of rural development as somehow separate from urban development. Unfortunately some even contrast agricultural and industrial development without appreciating that agriculture itself has a large industrial component essential to farming and to processing the products for use. The physical and biological environments and the cultural and economic opportunities of farm and urban people are intimately tied together. For national development both must be considered together for the benefit of all the people. Those having concern for agricultural development must take pains to clarify that part which is farming and relate it to the other large sectors of agriculture.

Modern farms must have the services of a town. Adam Smith wrote about 200 years ago,

> Compare the cultivation of the lands in the neighbourhood of any considerable town, with that of those which lie at some distance from it and you will easily satisfy yourself how much the country is benefitted by the commerce of the town.

Today in any area of well-developed agriculture the reciprocal benefits are even greater than in his time. Good highways or railways are necessary. Where much farming is highly mechanized, even air service for spare parts of machines should not be far away from the farms.

In their national planning it is important that the newly developing countries provide for many primary towns, secondary centers, and villages that can serve farming, forestry and other resource users. As we mentioned earlier such centers can provide schools and medical services for the trade area as well as for the people in the towns. Many existing villages can grow into such centers as agricultural development goes forward where fairly heavy transport can be available. It would be better for both town and country if most of the agricultural input and processing industries were located in towns reasonably near the farms. The workers will need to be trained, of course, wherever the industries are located. The same kinds of soil surveys used to advise farmers about the best systems of farming can also be used for selecting sites for towns, factories, and roadways.

And the great advantages of towns with space is that the children can have good places to play. Much of the crime in the overcrowded cities begins where children have nothing interesting to do.

Also the open space in a new town should exceed that needed for recreation. It should provide for new and unexpected changes. In town planning, full use should be made of known trends in population and the economy. Yet experience shows that one of the first principles of town planning is to realize that not all changes can be predicted. No one knows what will be invented even in the next decade. Such towns need also reserved spaces nearby that can be used for factories that may give off fumes and dust but with the minimum of pollution effect on the town and farms. The costs of land for building sites should be held down to prevent serious crowding.

Now, of course, those giving technical assistance should be reasonably sure that the farms and producing factories would be successful. Such factories employ people who need jobs. The income from these activities has an important multiplier effect, as Lord Keynes suggested (1965), on business and jobs in the trade area.

In any area in any country, success with town-and-country development depends on considerable understanding by the local people and their leadership. The developments offer opportunities for better living and income in countless areas in both the developed and the newly developing countries. I have seen enormous improvements from one key addition such as a trunk highway, a new supply of water for industrial use, the availability of electric power for both farms and local industries, or the discovery of a new mineral resource. But in most that have done well the spirit of cooperating throughout the local trade area counted heavily.

As with the development of only a new farm settlement or a new town, the process takes time. Capital must be had, for a start. Chances for good success are enormously enhanced by looking at the needs and opportunities of the whole trade area to capture the highly important interactions between the potential services of the town and the other people and the resources in the trade area.

The advantages in living in such a trade area with space and jobs are enormous, compared to over-crowded cities or isolated areas without schools and other services. As the time goes on, with success local savings and funds from both direct and in-heritance taxes can be available for continued development. Ing Jiřî Brousek (1971) has an excellent statement about this kind of town-and-country development in Czechoslovakia where farm and city people may both have adequate services in appropriate centers without segregation of either farm or city workers.

Governments must avoid regional unequality and rigid de-tailed plans for town-and-country planning. The people must take part. As they do, and as agriculture and commerce develop, the participating citizens will add considerable political stability to the country.

With inheritance of land and with considerable out-migration of people, however a farming area can be bled of funds. In the American Midwest a new migration pattern appeared during the period beginning about 1880. Many families were large. One son might stay on the farm and the others move to the cities. With the death of the parents four to six people might share in the in-heritance. Thus the one who stayed would need to pay the others for their shares[1] in order to have the farm. Many of the farms were nearly paid for each generation with the money going out of the area to distant cities. With good town-and-country planning, most people would not need to leave their community to get jobs, and more capital could be available for further development of the local trade area.

Yet local areas must guard against the possible bad influence of a large construction project of some sort that may suddenly bring in many workers and demands for services, such as schools and the usual municipal services. After 2 to 5 years, with the con-struction completed, say of a large dam, the people of the local area may find themselves with a great economic let-down and many unused service buildings—"ghost" towns they have been called. The same sort of thing can happen when a mine is exhausted with-out warning.

With the people understanding the situation in advance, good cooperative planning can avoid most such losses by having other enterprises to substitute or by moving more slowly.

---

1. In some countries laws now protect against splitting farms into units too small for successful operation.

Trade

Trade within the country and with other countries is essential to economic development above the subsistence level. Many of the ldc's, for example, may need to import some fertilizers and machines to get farming going. As time goes on local factories can be built, although even then some materials may still need to be imported. Where farm products will be the main source of income, emphasis should be placed on the industrial sectors of agriculture. But if those "infant" industries are too highly protected by import taxes, it will not be possible for the farmers to use the fertilizers and machines and development cannot go forward. This can happen even with naturally occurring fertilizers. At times, at least, Chile charged their own farmers much more for locally produced sodium nitrate than they sold it for in North America and Europe.

We have already referred to the export taxes on palm oil (or palm oil nuts) that keep farm prices low. The same has happened in Ghana with cacao beans.

More generally, in Europe and North America sugarbeets are grown for sugar rather than importing that from sugarcane from tropical areas. In my youth I heard this defended by "the need for a cultivated crop in the rotation to control weeds," and later as an assurance of supplies of sugar in the event of war. Now this protected crop is so entrenched in some temperate areas that few dare challenge it.

I have no intention to go into the many ramifications of the enormous problems of fair trade policies for the newly developing countries. Yet the problem needs study in each country. I should like to refer the reader to an excellent statement of some of the basic problems by Harry G. Johnson (1967).

Trade will be essential to most developing countries in order to import some of the supplies they need to get agricultural development going. They will need loans from international banks and loans and grants from individual governments. But eventually they will have to export in order to import. President Echeverria-Alvarez of Mexico summarizes well these needs for his country and efforts for both industrial and agricultural development and exports in *Ceres* (1971).

Some cultivators can export raw or partially finished products from mining, as well as from farming, forestry, and fishing. Yet they do not need to have all the capital from government. Private investment from other countries can help. Proper safeguards can be taken for major employment of local people. Such manufacturing can assist to make things needed within the country as well as for export. In the process local people have jobs and training. All of the developed countries have had much such help from private investors of other countries.

The main goal is to strive for those exports that have a chance to earn needed foreign exchange and to make jobs at home, including as much of the processing and packaging as possible. Many of the containers can be made in the country of origin of the crop or other product.

Certainly there are potentials for forestry in several of the newly developing countries. In fact considerable research and demonstration is needed in this area, to make importers aware of the potentials. The number of kinds of trees per hectare in tropical rain forests is higher than in forests of temperate regions. Many of the good species are not well known in trade. Because of their sparse distribution within the natural forest it is uneconomic to take out one or two species only. Research and demonstration in the Republic of Zaire have shown that many of the formerly unknown species of trees are very good for some uses required abroad, and that after cutting the forest can be replanted to the species most demanded abroad. Similar efforts are now being made in other tropical areas of natural mixed forest. The exports can be finished pulp, paper, plywood, lumber, and so on rather than simply logs, in order to keep jobs at home. Fortunately FAO and others are conducting research to make the tropical forests more useful to the world at large without waste or depletion. The prospects are excellent.

I repeat that some problems arising from early colonial development, religion, race, and social differences need to be overcome so that all citizens have full opportunities for education and jobs without discrimination either by persons or by products (Robert E. Baldwin, 1966). This is one of the great responsibilities of political leaders. Unhappily some are acting in the reverse direction. Development can be achieved only by using just means to achieve justice, to paraphrase again the late Prime Minister, Jawaharlal Nehru.

10

Aid for agricultural development in the less developed countries

Actually a very large number of national, regional, and private agencies are working directly or indirectly to hasten agricultural development. The work of the Ford, Rockefeller, and other foundations is generally well known. But there are many other smaller foundations and groups in various countries concerned with research or other aspects of the problem. No one can argue with the merit of cooperative approaches among all the groups and the bilateral government agencies to help make effective efforts of assisting people in the less developed countries to help themselves, but it is a large task.

Many of the "ups and downs" of foreign aid from its beginning by the developed countries after World War II are given in the *Reality of Foreign Aid* by Willard L. Thorp (1971), along with selected figures and many examples of financial support, transfer of capital, and political problems. He also discusses the role of private enterprise and trade.

The results of research and of experience in agricultural development have been exchanged among countries for many years, including several of the less developed countries. This was done through publications and in many other ways. France, Britain, The Netherlands, Belgium, and other countries established research stations, advisory agencies, and the like in their former dependencies.

Many agricultural scientists from North America and Europe visited the undeveloped areas looking for new plants, especially strains that would be hardy and resistant to diseases and insects, and for better breeding stock of animals. Many are concerned also with soil management, implements, and so on. While there, these travelling scientists talked with the local leaders and made suggestions. Plantation managers from developed countries made demonstrations through their own operations.

The scientists of the U.S. Department of Agriculture, the U.S. universities, and others from North America and Europe had been carrying on what we now call "technical assistance" for many decades before World War II. They both learned and taught. The U.S. Government sponsored research in agriculture in Latin America long before 1940.

Meetings of international scientific societies dealing with various aspects of agricultural science, technology, and trade were attended by scientists from many of the less developed countries.

The International Agricultural Institute at Rome promoted world-wide study. It furnished a forum for focusing attention on the needs and opportunities.[1] Unlike Athena, none of the national, international, or private agencies dealing with improvements of agriculture, the food supply, and nutrition sprang fully armed from the head of Zeus! A great many thoughtful people had been preparing the way for years.

THE FOOD AND AGRICULTURE ORGANIZATION OF THE UNITED NATIONS (FAO)

During World War II President Franklin D. Roosevelt, with the encouragement of several who had worked internationally in this area, invited the allied governments to send representatives to a

---

1. I had the opportunity to be a delegate for the United States at the 14th and last session of the general assembly of the Institute in 1938 and was Secretary of the Agricultural Committee of the organizational meeting of FAO at Quebec in 1945. Several people attended both of these sessions.

United Nations Conference on Food and Agriculture (1943) which was held at Hot Springs, Virginia (UN, 1943). An interim commission was set up to prepare for a permanent organization assembly with L. B. Pearson of Canada as chairman.

Thus the First Session of the Food and Agriculture Organization was convened in Quebec, Canada, 16 October 1945. Forty-two nations were represented. They agreed to help the agriculture in their own countries and also to cooperate to help each other (FAO, 1946). Now some 121 countries take part in this great multilateral effort toward agricultural development, increased food production, improved nutrition, fisheries, forestry, and closely related activities, for the people of the world. FAO was first located temporarily in Washington and since 1950 in Rome, Italy. The late Gove Hambidge covered much of the early history in *The Story of FAO* (1955).

The story of FAO

FAO, as the principal multilateral agency for technical assistance for agricultural development, serves as the world center in this field. FAO sponsors many important conferences and meetings, which include representatives from governments and other public and private agencies. It collects statistics from governments and gives advice both to newly developing countries and to others needing such assistance. FAO scientists and engineers help with appraisals for the international and regional banks, the UN Development Program, Special Fund, and others. The regular funds for FAO come from member governments and special contributions. The technical staff is recruited from many countries.

Yet the job of FAO is not easy. It has, of course, the kinds of problems of any large organization with a field force working within a single country, plus those resulting from differences of view of its many member governments of both developed and newly-developed countries, and the many necessary relationships with other UN agencies that deal to some extent with aspects of the world food and agricultural problems.

In order to see the whole international effort for development of the United Nations, and the background for it, the reader can get a general view from a fine little booklet by Vernon Duckworth-Barker entitled *Breakthrough to Tomorrow* (1970).

International consultative group

A new effort called an International Consultative Group on Agricultural Research looks promising. The Ford, Rockefeller, and other foundations take part as well as the International Bank for Reconstruction and Development (IBRD), FAO, and the United Nations Development Program (UNDP). Others are the International Development Research Center of Canada and the governments of the United States, West Germany, Denmark, Sweden, The Netherlands, the United Kingdom, and France. Probably other governments and regional banks will take part. An international effort of this sort can do a lot to set priorities and avoid unnecessary duplication. If a research effort has produced dependable results on representative kinds of soil in one country, the basic work does not need to be repeated on the same kind of soil elsewhere. But as we have emphasized before, the combinations of practices used must be adapted to the local skills of the people and their needs and services.

United States assistance abroad

Almost immediately after World War II, a special agency, United Nations Relief and Rehabilitation Administration (UNRRA), was established to which the United States gave considerable assistance. This helped greatly with the immediate critical problems of overcoming the war devastation in Europe.

After much preliminary work and study, Secretary of State George C. Marshall gave his famous speech at Harvard University on 5 June 1947, offering assistance for European recovery. After hesitation, the Soviet Union decided not to take part. The Economic Cooperation Administration, ECA, reporting directly to the President, administered the work. But the name "Marshall Plan" also stuck.[2]

I recall being out in the bush in Zaire (Belgian Congo) at the time Secretary Marshall made his speech, but, of course, I could not answer any of the burning questions of my Belgian associates. Later in 1948 and 1950 I was in Western Europe. The effects of the Marshall Plan were remarkable. Many have said that the Marshall Plan was probably the most important peace-time effort abroad of the United States. Yet the very success of the Marshall Plan misled many Americans. To them it seemed easy to make enormous progress in foreign assistance with United States funds and staff.

First of all the whole program of the Marshall Plan was quietly and carefully studied in advance by a committee that included people very well informed on one or more of the European countries.[3] Also a high proportion of the staff and consultants were well qualified and dedicated people—scientists, engineers, managers, and so on—from government, universities, and the private sector.

Equally important, Western Europe had large numbers of skilled managers, scientists, engineers, and technicians. They needed some funds, encouragement, and cooperative planning to get going. Which they did. The advanced planning and the combined efforts led to such a great success that technical assistance abroad looked easy!

Probably few other technical assistance projects were so well planned by people highly informed both in their fields *and* about the countries. Many failed to realize fully that much fewer experienced and competent managers, scientists, engineers, and technicians would be found in the less developed countries. Further many of the less developed countries were already, or soon would be in turmoil over independence or political strife within their countries, or both.

Point Four

Early in 1949 President Truman gave a strong speech on foreign relations in the inaugural address of his second term. He emphasized both political and economic recovery in the world. He

2. In 1951 a Technical Cooperation Administration (TCA) was established. Then the Mutual Security Agency (MSA) was created in 1951, especially for military aid. In 1953 MSA and TCA were transferred to the Foreign Operations Administration (FOA). (Naturally this caused a bit of confusion with FAO in Rome.)

In 1955 the International Cooperation Administration (ICA took over the functions of FOA; and in 1961 The Agency for International Development (AID) was organized. In the process, some of the financial and loan agencies were separated from technical assistance and the Public Law 480 program (Food for Peace) was separate. (See Willard L. Thorp, 1971.)

3. See Dean Acheson's *Present at the Creation* (1969) and especially George F. Kennan's *Memoirs: 1925-50* (1967).

listed the "four main courses of action," of which the fourth dealt with technical assistance. He said "Fourth, we must embark on a bold new program for making the benefits of our scientific advances and industrial progress available for the improvement and growth of underdeveloped areas."

Thus it was that "Point Four" remained a general name for US technical assistance for several years. Added to the new political problems resulting from independence or efforts toward it, the United States bilateral technical assistance programs came at a time when many in the less developed countries had high expectations for rapid improvements in their status and living standards.

Some of the people in these countries knew about standards of living in most of Western Europe and North America and knew that their misery was not necessary. But few knew how to go about their own agricultural development nor did they understand the enormous amount of study and especially hard work that would be required.

Many programs

Besides the FAO and the US bilateral technical assistance, other developed countries, also have given help in agricultural development to the newly developing countries. Both France and Britain have given aid to their former dependencies. Then too, the British Commonwealth had a Colombo Plan for mutual assistance.

The Soviet Union gave loans at low rates of interest. With United States help great improvements in farming were made in Formosa (Taiwan) and, in turn, some of the people there gave help in other countries. Assistance by China has been more recent.

Some program problems

Even with the many governmental, international, and private agencies—and only part of them have been mentioned—trying to help people in the less developed countries, the total accomplishment in agricultural development has been much less than many had hoped for. Several factors complicate the working together of sovereign countries.

As we pointed out, the Marshall Plan in Europe had an excellent preparation and good staff. After the emphasis was switched to the less developed countries the general word about the "Point Four" program was "temporary." Much of the staff that worked in Europe returned home to responsible positions in government, universities, and the private sector. It became difficult to recruit the kind of highly qualified, dedicated, and broadly educated people needed for technical assistance.

The advanced preparation for the conditions in each of the countries assisted was minimal compared to that for the Marshall Plan. Before going abroad some assistance workers knew little about the countries to which they were assigned.

Then too, military aid and technical assistance did not always ride well together. Some political leaders were more interested in military aid than in technical assistance and finance to get agricultural development going. Despite the fact that most developed countries of the world got their start out of a good agriculture, farming still has a low status with many government leaders.

Naturally some political problems arose. One Minister of State from a little country in Africa told me, "I simply cannot get any United States aid because we have almost no Communists." I tried to reassure him that this would not be a factor, but with unknown success.

In the United States there arose a special problem called "McCarthyism," especially during the late 1940's and early 1950's resulting from investigations by committees in both houses of the Congress looking for disloyal Americans. The intentions and integrity of many able and loyal Americans were questioned publicly, if not directly, by innuendo. Even some upright officials, especially the educated ones, were "accused" while carrying on their duties abroad. [See Dean Acheson (1969) who was himself harassed!] This harmed the functioning of the U.S. Government considerably and many able people avoided positions in the government, especially those having to do with foreign affairs.

With so many agencies of the UN, individual governments, and private agencies helping countries, many of the less developed countries have trouble with priorities. Despite official declarations of intent of cooperation by aid-giving groups, not all countries actually seek cooperation with one another. Most cooperation among them depends mainly on the skill and good will of the individuals on the ground. Some choose to work together and keep each other informed, and others do not.

Many countries and agencies had projects in India, for example, when I was there. It took a lot of the time of people in the Government of India just to guide them about and talk with them. Further, for any project to be effective, qualified Indians had also to be assigned to it. This required the services of a considerable fraction of the qualified people in many aspects of the arts, sciences, engineering, and technology.

One of the greatest problems, in my view, has been the search for "reasons" why people of the developed countries should help those of the less developed countries. Many reasons have been advanced by governments of wealthy countries "to explain" to their citizens how they themselves will benefit from giving aid. Some say that if all countries were well off, wars would be reduced. Yet taxes for military preparations increase. It is said "as we help the less developed countries, people in our countries will benefit from the increased trade." And so on.

Actually the only argument that holds is the one of compassionate moral duty. As Gunnar Myrdal explained recently, preference should be given to those undeveloped areas where the people and their leadership are committed to internal reforms for justice (Myrdal, 1972).

Multilateral versus bilateral technical assistance

This question is being argued a great deal in the United States and elsewhere. Some say that FAO should be more generously supported by member governments and handle essentially all the technical assistance in agricultural development and related fields, with loans handled by the international and regional finance agencies. Except for the critical matter of financial support of FAO by governments, I should favor this plan. The government of the United States and some other governments have not been especially generous with their support of FAO. Nor will they unless more adequate steps are taken by the individual governments to explain to their people, including legislators, the purposes and accomplishments of FAO.

The United States did sponsor an FAO World Food Congress in 1963. Recently the Canada Department of Agriculture issued a fine bulletin on *Canada and FAO* (1971). Still the people in the developed countries need to know more about this great effort.

As time has gone on more American-trained people have gone

abroad to help the people in the less developed countries and more of their young people have been trained in U.S. universities. On the whole, communications are improving. Now it is possible to find many examples of both multilateral and bilateral technical assistance. And cooperation among the various agencies and governments appears to be improving.

Requirements of those carrying out technical assistance

As we have suggested, the basic principles of agricultural development vary little from place to place or with time. But the problems of fitting the people, with their skills, goals, and social systems, to the kinds of soil and other resources in order to achieve successful agricultural systems vary widely indeed, depending on the social systems, resources, customs, and political organizations within individual countries and parts of countries.

As discussed a bit earlier, neither the very narrowly trained specialist nor the ill-prepared generalist can be expected to be fully helpful, except that the specialist may perform certain useful functions under a qualified leader, and at the same time can be trained in how his function relates to the whole.

What is really needed are dedicated people specialized in depth in some one field with interdisciplinary reading and experience in the other natural and social sciences. At least some knowledge of religious and cultural systems, along with integrity and good manners, are essential to gain the respect and confidence of other kinds of people whether they be simple cultivators in the "bush" or prime ministers. Mere sympathy for people, however, is not nearly enough.

Very high on the list of qualifications is willingness to work. Perhaps the most embarrassing question that many ministers of state in newly developing countries have asked me in private conferences, runs about like this "I have been in the United States, I know how hard your people work. Why will they not work when they come to my country"? Of course, many in technical assistance abroad do work hard. No doubt the drive to work to the best of their ability is the most important lesson for many to learn. All good agricultural systems depend on it, not only on farms but also in all the other sectors. Examples set by those furnishing technical assistance can be influential, one way or the other. The best advisor also does all he can reasonably to help train local advisors to carry on and to be able to replace the advisor himself after a good start with a needed program that the people want to continue. The example should not be spoiled by insisting that everything needs to be done in one particular way.

Differences in competence have produced highly contrasting results. The Gezira Scheme in the Sudan, to produce cotton with irrigation on a very clayey soil, called Tropical Black Clay or Vertisols, was excellent. The British started the research for it in the early 1930's (Tothill, 1948). Another amazing effort was made by A. B. Stewart of Aberdeen, Scotland in India, during 1946. In less than a year of field study he developed a basic report for the Council of Agricultural Research, New Delhi, on the interpretation of their own research and experience with fertilizers, together with recommendations of soil and fertilizer research to improve yields of different crops growing on widely contrasting kinds of soil with unlike climatic conditions (A. B. Stewart, 1947).

On the other hand the Groundnut Scheme in East Africa was begun without even exploratory scientific research. Few examples could be found of such an exceptionally costly lesson on the need for soil and other research examinations in advance by scientists (Wood, 1950). Let us hope that this kind of unscientific approach is behind us.

11

The world outlook for agricultural development

Let us summarize briefly where we have come and some of the opportunities and problems ahead.

The world has not made the progress toward agricultural development that many had hoped for in Quebec in 1945. We know now that we were thinking in terms of decades whereas we should have taken a much longer view. Still the food supply is increasing substantially, by using only about one-half, or a bit less, of the potentially arable soils of the world. The increases in food have been mainly, although not wholly, on soils that were in use about 1945.

My old friend Dr. Vladimir Ignatieff, affectionately known as "The Major," "Mr. *Efficient Use*" (1958), or "Mr. FAO," was at the organizational meeting of FAO in 1945 and was a senior member of the FAO staff until his retirement late in 1969 to his home in Canada. We have had many days of discussion in Washington, Rome, our homes, and elsewhere over these years. We have agreed that at the start we failed to emphasize nearly enough the work—the hard work—required by all of us, and especially by the people on the land and in their governments, to achieve reasonably effective agricultural development. As emphasized earlier many were deceived about the ease and spread of development by the great success of the Marshall Plan in Western Europe. It looked easy. Then too, the improvement of farming in the Soviet Union after recovery from the horrible devastation and confusion of World War II was remarkable. Both Western Europe and the Soviet Union had many skilled people. But the job was greater in most of the less developed countries. Much hard work is still ahead.

The great strides forward in health and in food supply have reduced the death rates. Now there are more of us to feed. Yet efforts toward lower birth rates are progressing as other forms of social security besides "many sons" have developed.

Many countries have gone through the inevitable confusion of independence and the formation of new governments. Of course, these governments are not fully experienced. Yet certainly we are justified in being hopeful of many of them despite certian pulling and hauling by other countries and blocs. The end of this is hardly in sight. Some people in parts of Africa and elsewhere are still held in semibondage. Their outlook is unpleasant, to say nothing of their lack of incentives for agricultural development. Progress will be uneven.

However a lot has been learned. We have had some remarkable successes here and there and some costly failures, such as schemes for production, land reform, and the like having been planned without the facts about soils, water, climate, and transport essential for success. These failures made ineffective some scarce funds. Yet many have learned how to get the facts and go ahead with such developments. And we cannot allow ourselves to be too discouraged with a few more failures from lack of competent staff somewhere along the line. Many excellent people have been trained, partly through experience, to use both the natural and social sciences, and technology, in a reasonable mix. May we have many more in the next decade.

Wars and revolutions are going on now while I write—completely unnecessary killing and destruction of the works that people have built. Probably we must expect some more. Then too, we must expect destructive natural disasters, as in the past, such as great floods, tidal waves, severe droughts, unusual cold, earthquakes, and volcanic eruptions. I mention these because we already lack the minimum infrastructure for agricultural development without either man-made or natural disaster. So, of course, we lack storage and transport facilities for reserve supplies of food to meet disasters.

Yet on the other side, governments and people have shown that they can be generous in helping one another. Substantial sums have been supplied to help the less developed countries get started on their way. These need to be increased for well planned efforts in research, technical assistance, and for development of the infrastructure, including roads and dams to have water for hydroelectric power, irrigation, and local industries.

In many places now the wastes and losses of farm products from not using these great potentials for power and refrigeration are enormous. It seems curious that some groups now urging more care of the outdoor environment, including avoidance of air and water pollution, also actually oppose dams for hydroelectric power, which can be produced economically at good sites without polluting either the air or the water.

Statistical data are inadequate

Handicaps to knowing where we stand in the world in agricultural development are the inadequate figures on food and other farm production, food consumption, and the number of people working in the principal sectors of agriculture. Most tables listing those working in agriculture appear to give mainly the numbers working on farms, not including those making the essential inputs, processing the farm products, and so on. Yet we know that these workers are also essential to agricultural systems. In the more primitive areas the numbers of workers on farms and in farm villages are high. In the developed countries the manufacturing of inputs, processing, and the like are mainly in towns and cities so the numbers on farms are much lower.

FAO no doubt has done as well as possible in its compilation called *The State of Food and Agriculture.* Those compilations for recent years also have useful summaries of trends in trade, technology, and other important aspects of agricultural development. Those for 1969, 1970, and 1971 are helpful in the prediction of important trends (FAO, 1969, 1970, 1971). The trends for several years and the outlook for international agricultural adjustments to 1980 were given in an excellent paper by Oris V. Wells, former Deputy Director-General of FAO (Wells, 1971).

Member countries mainly furnish the figures for *The State of Food and Agriculture to FAO.* For countries that do not support FAO only partial estimates can be given. Probably the figures for farm products that enter foreign trade, and those traded within the developed member countries are reasonable.

The United States contributed a useful summary of the situation, *The World Food Problem* by the President's Science Advisory Committee (1967). Another important piece of work is the *Provisional Indicative World Plan for Agricultural Development* (FAO, 1970), published in two large volumes. Trends and projections to 1985 are given on many activities directly related to food and consumption of food, trade, and to on-farm agricultural development, including fisheries and forestry. The development of essential infrastructure and industries in the agricultural sector, are treated in far less detail. These books are very useful to many seeking a general view of broad trends and emerging problems and opportunities, especially to those accustomed to normal writing in bureaucratese.

For a brief statement, the United Nations has issued a small terse bulletin on *International Development Strategy* (1970) projecting to about 1981. This includes most broad aspects of the general subject of development.

I have been in many local village markets in Asia, and especially in Africa, where little money, if any, was used in trade, and where certainly no one was recording the transactions (Figure 84).

I suspect that the world figures on the so-called "protective" foods are considerably low, certainly as related to rural diets. Many have pointed out, for example, the lack of minerals, proteins, and vitamins in cassava (manioc, tapioca). This is true of the roots,

FIGURE 84

This is one of many thousands of local village trading centers without records.

which do go to the cities. But a visitor to the homes of the cultivators may see them prepare the tops, reasonably rich in vitamins
and protein for greens.

I recall reading about nutrition in Mexico before going to
Mexico City and the nearby area in 1942. I had read that the
people's diets were inadequate in ascorbic acid, vitamin C. Yet I
saw very little evidence of vitamin C deficiency. The reason was
that no one had counted the widely used little red peppers, so rich
in ascorbic acid!

Local people get fish and game which are not counted. The
same is true of fruits and nuts of wild or local trees and palms and
the products of kitchen gardens. This is partly responsible for some
overstatements of low food in farming areas during normal seasons.
It was not all counted.

The fact is we lack reasonable data in many areas of the kinds
and amounts of food people eat and where they get it. This can be
done more easily in towns and cities than for the country. In
many of the poorly developed country areas, it would be helpful
for a scientist who understands soils and the principles of sampling,
aided by local informed people, to plot reasonably homogeneous
areas on maps of roughly 50 to 250 square kilometers, or perhaps
even considerably more in thinly populated areas. Then one or
more representative farm villages could be used as samples. If arrangements could be made for a person knowledgeable about food,
nutrition, and sampling to stay in the sample village for a year to
keep account of what people eat and where it comes from we could
begin to get some useful data. In some villages account would
need to be taken of unusual seasons that are likely to cause poor
crops.

Then too, of course, some people put a bit too much confidence in published figures for trade in cereals, such as barley,
wheat, maize, and rice. Even worse, some have gone so far as to
equate these with "food"! Actually many people depend mainly
on locally grown bananas, cassava, taro, potatoes, sweetpotatoes,
and the like. Of course, some of the published errors could have
been avoided with only a little knowledge of the countries.

The advisor must go to the farms and villages

My own experience has been that in most countries the capital city
is not the best place to find out much about current farming. First

of all, the statistics are not fully reliable but they are given out. As I have urged, to learn much of value about yields and all the rest one must go to the farms, villages, and country towns. Thus for part of the time one needs a good interpreter. I do not mean to imply that people in the agricultural ministries intend to mislead one, but they often take their statistics too seriously. Officials living in capital cities do not always get onto representative farms and see how the cultivators actually carry on their work and trade. One must also visit with them to find out what their problems are —at least *their* views of what they are.

In one country, for example, I recall asking the official responsible for olives "what proportion of your olives are irrigated?" He replied: "None. We never irrigate olives in this country." Yet on my tour directly afterward I saw many hundreds of hectares of well-irrigated olives, which shows that cultivators had considerable skill with the system. Because of such experiences, I have made a firm rule that on my visits to see the agriculture of a country that I give no written report to the Ministry without opportunities to to visit the cultivators and to talk with them (Figure 85).

FIGURE 85

The author poses with cocoa farmers after a conference in Ghana.

Now, of course, one needs also to talk with the people in the ministries of Agriculture and Finance, both before the field tours and afterward. These people have their problems too. It is unproductive to be thinking in detail about projects or schemes that could not possibly be financed or staffed. Just after the failure of the Groundnut Scheme I recall the head of a distant State in Africa saying to me "you report the truth just as you see it, but if you recommend any schemes, *please* keep them small to begin with."

Changes in the environment

To have good farming, people have learned that many changes in the natural system must be made. In large parts of the world trees must be removed, runoff controlled, and plant nutrients added. Lately many people speak of only "protecting" the environment. Generally speaking, with some outstanding exceptions where unresponsive soils were selected, the change to adapted farming systems has improved the environment. The exceptions were due partly to changes in climate in the very early days of man's appearance, to migrants not being able to judge the soils of a strange area, and to wars and serious economic difficulties.

With more people needing food and the drive toward efficiency with high yields, chemical fertilizers have had to be used. Where these are used prudently there is little evidence of significant harm in relation to the enormous benefits. On the increased use of pesticides to protect man himself, crops, and animals, the current situation is less clear. DDT used to control insects, including mosquitoes that carry malaria, has been a whipping boy for many ecologists. Yet I have seen no firm evidence that *ordinary amounts* from *normal* applications injure either people or animals. Probably most harm that has resulted from the pesticides occurred because of the lack of control of factory wastes and some from the use of excessive amounts. Then too, other factory wastes have caused some serious problems of water pollution, such as the disposal of mercury from the manufacture of paper and other industrial products that found its way into fish (Nelson et al., 1971). For the United States the unclear situation is set forth in the *Report of the Secretary's Commission on Pesticides and their Relationship to Environmental Health* (U.S. Dep. of Health, Education, and Welfare, 1969).

Recently "scare" books and papers on the environment have been numerous. Some are helpful but many are exaggerated with too little regard for the scientific facts. For example, I've read such statements as, "The tropical rain forests should not be cleared for farming because it would take all the oxygen out of the air to oxidize the iron in the soil"! And also that "all the soils in the Amazon Basin would change to hard laterite if the forest were cleared." Possibly some laterite would be hardened if land were foolishly cleared and plowed without adequate surveys in advance— possibly as much as 4 or 5% of the surface in places. Actually some of the soils with the highest potential for farming are under little— used tropical rain forest, especially in Central Africa. Of course, it will take work to prepare them for use.

Many complain about the nitrogen in rivers rising in mountains with good soils for trees where thunderstorms are common. Of course, there is nitrogen; it helped make the old Nile Valley so fertile long before nitrogen fertilizer was available or even known. Recently someone raised a question, seriously, about the uranium in some phosphatic fertilizers! The competition for "scare" stories is keen!

Others complain about the sediment in rivers flowing through farming areas. Considerable sediment does come off farm land with soils wrongly selected for such use, or badly managed, that finds its way into rivers. But as much or more of it comes from a few critical sources not in farms but along the "cut banks" of meandering streams during high flows and from deep gullies in the high country. The word "Missouri" of the Missouri River means "big muddy" in the local Indian language. Of course, the Mississippi Delta is no doubt growing. Without farming it might have grown more slowly. Possibly a large share of the sediment in the Mississippi Delta comes from poor runoff-control practices and erosion in the towns and cities, and from poorly placed roads and factories. No one knows quantitatively.

Some people complain "on principle" against clear-cutting (cutting all the trees) in forestry. Actually this can be an excellent practice if the soils are stable and current trees are a poor species. Huge forest fires during dry periods are commonly followed by what are called "weed trees" or poor trees. Some of the best adapted species cannot grow well in the shade. Thus the best solution may be to clear cut and plant good species. On very hilly, unstable, or erosive soils, of course, clear cutting is not a good practice.

There have also been many questions raised about nitrates in soils and water. As we have seen, nature has many sources of nitrogen including thunderstorms and nitrogen-fixing organisms, as well as fertilizer. The whole situation is still not completely clear, despite some obvious exaggerations. Frank G. Viets and Richard H. Hageman have made a searching inquiry about what is known in a recent helpful bulletin (Viets and Hageman, 1971).

It is true that many concentrated animal and poultry feeding operations result in huge piles of manure. Formerly such manure was carried to fields for its fertilizing value until the price of chemical fertilizers became much lower. Preventive practices do need to be taken, perhaps by some scheme of licenses, to prevent water soaking through these manure piles and going directly into streams. Also in places more effective measures are needed for controlling sewage-plant effluents and drainage waters from some areas of pervious soils that are irrigated and heavily fertilized. The sulphur from burning sulphur-high oil and coal can lead to air pollution that has bad influences on plants.

Perhaps these are enough examples to illustrate the fact that we must be careful in our advice to the newly developing countries. The combinations of practices needed to enhance all four environments—cultural, biological, physical, and economic—are very different in different places. Few if any generalizations fit all places. Yet with good planning and *work* all these environments can be improved and can support both good farming and furnish excellent forests, with the essential infrastructure.

Outlook in summary

In trying to put myself back to the time of the great discoveries, certainly I should have been less optimistic than now. So many of the early immigrants to North America had enormous handicaps. They had so much land to clear, even to get started. Building on the experience of Western Europe and of the Indians already in North America, they did develop fairly rewarding farming systems on soils of low productivity when first cleared. Fortunately, science and technology came along to help, slowly for a time but with steadily greater effects, with aid from the books of Arthur Young and others near the end of the eighteenth century. He had a good influence in the United States on several, including George Washington—the first President (Washington, 1803). The first really great book on soils written in the United States appeared in 1832—

Edmund Ruffin's first edition (as a book) of *An Essay on Calcareous Manures.*

The U.S. Department of Agriculture and the land-grant colleges of agriculture were initiated in 1862. After that more research got underway, especially after national cost-sharing for research with the states in 1887. This was followed by national, state, and local cost-sharing for an Extension Service in the early part of this century. But before that some efforts were made at the request of President Thomas Jefferson in the early Patent Office. In fact the continuing series of the U.S. Department of Agriculture Yearbooks was begun in the Patent Office. In addition several privately made inventions in Europe and North America were especially useful in mechanization of arduous farm operations.

The long history of the development of fertilizers is also very impressive during the latter part of the nineteenth century and even now. During both World War I and World War II chemical processes for manufacturing nitrogen fertilizer were improved and cheapened. Phosphatic fertilizers were improved, especially by the Tennessee Valley Authority after 1935, and also by some commercial producers. Soil tests for the amounts to use for specific soil-plant combinations were available to nearly all farmers in the United States by 1945. Until very recent years costs of fertilizer to farmers had been still further reduced. All of this took a long time and much hard work.

Fortunately, the newly developing countries can use a high proportion of the basic principles of agricultural science and technology developed in the eighteenth, nineteenth, and twentieth centuries. But, of course, these principles must be adapted to the local kinds of soil, forests, crops, climate, and people. Yet the problem looks to me to be less difficult than that of the very early settlers in North America.

There is no substance to the irresponsible statements that the world lacks potentially arable soils. It is true that the essential infrastructure for agriculture is lacking in large areas and it will take hard work to build it, but certainly it is not so difficult as 150 years ago. Part of the job can be done in areas of high population by labor-intensive methods without importing labor in the form of giant machines made elsewhere. But some heavy machines are needed for land clearing.

This in no way argues against continued efforts toward improved health and all reasonable measures to reduce birth rates, especially by substituting other forms of social security in old age for "many sons." Nor does it argue against adapted combinations of measures to protect people and their basic resources from deterioration, not only from the way they were in nature but from what they could be with proper management.

From what I have learned of modern social and natural sciences and of the potential for technology, town-and-country development, and the basic good will and willingness to work of most healthy people, I am optimistic for the good agricultural development the people of the world need, *if* we are able to go about it by means that are just to all. The world has ample resources to provide the agricultural products people will need for a very long time.

Many qualified people have already had useful and successful experience helping people and their leaders with agricultural development in several of the developing countries. FAO and all governments need more such competent people. So do we all.

Despite the fact that a tiny few oppose technology, I feel optimistic. But, of course, the concepts of peace, justice, and appreciation of other people need stronger support for the appropriate sciences and technologies to be learned and applied to the abundant resources of the world.

All of us concerned in this urgent effort have a lot of hard work to do.

Glossary

ACID SULPHATE SOILS. Soils mainly in moist areas near the sea, with considerable organic matter, that become extremely acid when drained. (See text.)

ALBOLLS (1975). Mollisols that have gray horizons below the dark surface soil, are underlain by clayey horizons, and are commonly wet near the surface during winter and spring.

ALFISOLS (1975). Soils with horizons having accumulations of silicate clays beneath that may or may not be underlain by hardpans.

ALKALINE SOIL. A soil that is alkaline throughout most or all of the rooting zone, with pH higher than 7.3.

ALLUVIAL SOILS. Soils developing from recently deposited alluvium.

ALLUVIUM. Sand, mud, and other deposits by streams.

ANDEPTS (1975). Inceptisols that are developed mainly from old or recent volcanic ash and cinders.

AQUENTS (1975). Entisols that are wet, as in tidal marshes and near lakes and streams.

AQUEPTS (1975). Inceptisols that are wet. They may be found in cold or tropical climates.

AQUODS (1975). Spodosols in wet or moist places with fluctuating ground water. Formerly called Ground-Water Podzols. Can be found in tropical as well as cool climates.

AQUULTS (1975). Grayish or olive-colored Ultisols with ground water near the surface in winter and spring of warm temperate regions.

ARGIDS (1975). Aridisols that have clayey horizons beneath the surface.

ARIDIC. In aridic climates the soils are dry throughout more than one-half the time.

1. Dates in parentheses refer to date of 1938 soil classification used in Figure 8 (USDA, 1938) and 1975, the approximate date of the *Soil Taxonomy*, USDA (in press), with a few other dates.

ARIDISOLS (1975). Generally soils of arid regions with scanty vegetation. Some respond well with irrigation.

BLACK COTTON SOILS. Former name for Vertisols (1975).

BOG SOILS (1938). Soils developed from peat, usually wet unless drained.

BORALFS (1975). An Alfisol (1975) in cold moist regions.

BOROLLS (1975). Mollisols with a black surface layer over a brown horizon and a fairly low annual temperature.

BROWN FOREST SOILS (1938). Soils with dark brown horizons, rich in humus, and grading down into lighter colored soil. They are developed mainly under deciduous forest in temperate or cool humid regions.

CATENA, SOIL. A group of soils within a soil zone from similar parent earthy materials but with unlike characteristics owing to differences in soil slope and drainage.

CHERNOZEM SOILS (1938). They have dark colored, nearly black surface soils with lighter colored horizons containing calcium carbonate beneath. They are developed in cool subhumid regions under tall or mixed grasses. The term is Russian for "black earth."

CHESTNUT SOILS (1938). Soils with dark-brown surface horizons with lighter colored soil beneath. They are developed in a somewhat drier climate than Chernozem soils.

CLIMATIC MOORS. A term used for hilly, peaty soils kept nearly continuously moist by mist and rain.

COMPOST. A mass of well decomposed organic matter made from plants. Some add manure and a little fertilizer, especially nitrogen. The whole is kept moist and commonly turned once or twice during a season before use.

CRYIC SOILS (1975). Soils in very cold regions.

DEGRADED CHERNOZEM SOILS (1938). Soils that have very dark thin surface horizons, with light colored horizons beneath; usually former Chernozem soils that have been changed by the coming of a more moist climate and invasions of trees.

DESERT SOILS. Usually light-colored soils of deserts with little vegetation.

ENTISOLS (1975). Soils dominated by earthy matter not yet modified much by soil forming processes.

GILGAI. The characteristic microrelief of Vertisols (Black Cotton soils) due to the high swelling and shrinking with wetting and drying. (A word from the early native people of Australia.)

GLEI OR GLEY. Soil material that is mottled gray or bleached under poor drainage in the presence of organic matter.

GRAY-BROWN PODZOLIC SOILS (1938). Soils with relatively thin organic covering in nature, and a thin organic-mineral horizon over a grayish horizon, which is underlain by a clayey horizon. Developed in humid temperate regions under mainly deciduous forest.

HALF-BOG SOILS (1938). Soils developed from shallow layers of peat over lighter colored glei horizons, usually moist.

HISTOSOLS (1975). Soils developed mainly from organic materials, especially accumulations of peat.

HORIZON, SOIL. A distinctive layer of soil roughly parallel to the surface, except for tongues and minor irregularities, with distinct characteristics produced by the active processes of soil formation.

HUMODS (1975). Well drained Spodosols with only thin surface gray layers and nearly black, reddish horizons beneath. Formerly included with Podzols (1938).

HUMULTS (1975). Well drained Ultisols in mid- or low-latitudes with considerable organic matter in the surface soil. The vegetation is dominantly trees.

HYDROMORPHIC SOILS. A general term for soils strongly influenced by water at or near the surface much of the time.

INCEPTISOLS (1975). Soils with moisture more than half the year but without concentrations of clay in lower horizons.

LATERITE. Highly weathered clayey material rich in iron and allumina that hardens irreversibly when dried, now called also plinthite. (See text.)

LEACHING. Removal of materials in solution by water.

LITHOSOLS (1938). Soils with thin, and usually irregular, horizons over rock.

MATURE SOIL. A soil with well-developed horizons in near equilibrium with its environment.

MOLLISOLS (1975). Soils with very dark brown-to-black surface horizons. They form under grass in regions of moderate dry periods of varying lengths.

MUCK. Fairly well decomposed organic material relatively high in mineral content, dark in color, and developed from peat formed under conditions of poor drainage.

NORMAL SOIL. A soil with characteristics in near equilibrium with its environment, developed under good but not excessive drainage and expressing the full effects of climate and living matter.

OCHREPTS (1975). Light-colored, freely drained Inceptisols on moderate slopes in temperate to cool climates.

ORTHENTS (1975). Entisols on recent erosional surfaces and eolian deposits. They are found in most climates.

ORTHODS (1975). The more commonly found Spodosols of North America and Europe, formerly called Podzols. They are developed under coniferous or mixed coniferous and deciduous trees. Under cultivation the gray surface soil and dark lower horizons are mixed.

ORTHOX (1975). Oxisols with a short or no dry season.

OXISOLS (1975). Soils with well weathered minerals. The clay is weakly active and commonly decreases with depth. They are soils of the humid tropics and subtropics.

PALEOSOLS. Very old soils preserved by subsequent coverings of loess or other fine earthy materials.

PEDON. The smallest area of a soil that can be used for describing the soil horizons and their lateral variations and for sampling. Commonly these vary in area from 1 to 10 square meters.

pH. A notation to indicate weak alkalinity or acidity, as in soils. A pH of 7.0 indicates neutrality, higher values alkalinity and lower ones acidity.

PLINTHITE (1975). A more specific term for "laterite," since the older term laterite was loosely used by some. (See text.)

PODZOL SOILS (1938). Soils having an organic mat in nature and a gray leached horizon underlain by a dark brown horizon, commonly partly cemented.

PODZOLIC SOIL (1938). A leached soil with some of the properties of the Podzol.

POLYPEDON. A mappable area of similar pedons, commonly called an individual soil. These are the principal mapping units in detailed soil surveys.

PRAIRIE SOILS (1938). Soils of temperate, relatively humid regions with very dark brown to black surface horizons developed under grasses.

PROFILE, SOIL. A vertical section of a soil through all its horizons down into the material beneath the soil.

PSAMMENTS (1975). Entisols in well sorted sands of sand dunes or cover sands.

RED DESERT SOILS (1938). Red soils of presently or formerly hot deserts.

REDDISH CHESTNUT SOILS (1938). Soils similar to Chestnut soils but in warmer climates.

RED-YELLOW PODZOLIC SOILS (1938). Red and yellow leached soils with weak or strong clayey horizons developed under mixed forest in warm temperate moist climates, including some in tropical regions.

REGOSOLS (1938). Soils developed mainly from soft or unconsolidated rock, with or without thin coverings of true soil.

RENDOLLS (1975). Mollisols of humid regions that were formed mainly under forests from very limy parent materials, such as chalk or highly calcareous glacial drift. (Formerly—1938—called Rendzina soils.)

SALINE SOILS (1938). Soils with more than 0.2% soluble salts, but not highly alkaline. Also called Solonchak soils.

SALTY SOILS. See saline soils.

SIEROZEM SOILS (1938). Light-colored, little leached soils within near desert climates.

SOIL ASSOCIATION. A grouping of geographically associated but unlike kinds of polypedons that cannot be shown separately on soil maps at the scale being used. Such associations are thus used as mapping units.

SOIL BLOWING. The movement of soil material from soils by wind.

SOIL EROSION. The movement downslope of soil or soil material by water.

SOLONCHAK SOILS. See saline soils.

SOLUM. The upper part of the soil profile, above the parent material, in which the processes of soil formation and most root growth takes place.

SPODOSOLS (1975). These soils have gray surface soils under an organic mat in nature and have a lower horizon of reddish-brown or nearly black amorphous materials. They are commonly found in cool, moist climates, but some can be seen in ill-drained sandy deposits in the tropics and subtropics. Formerly called Podzols.

SURFACE SOIL. The upper part of cultivated soils commonly stirred in plowing or by other tillage implements.

TERRA ROSSA. Generally, red soils from calcareous rocks developed under thin forest in the kind of climate around the Mediterranean Sea.

TOP SOIL. (1) A general term for surface soils. The term is vague and can be defined precisely as to depth and productivity only in respect to specific kinds of soil. (2) Also used for presumably mellow, fertile soil for top dressing lawns and gardens.

TROPEPTS (1975). Brown to reddish Inceptisols mainly in tropical or near tropical areas.

TROPICAL BLACK CLAYS. Name in older literature for Vertisols (1975).

TUNDRA SOILS. Generally, soils with a tough fibrous mat on the surface underlain by a few centimeters or more of humus-rich soil. In depth this gives way to gray or mottled soil over ever-frozen substrata.

UDALFS (1975). Alfisols that are rarely dry and in humid temperate climates.

UDERTS (1975). These are Vertisols of fairly humid climates and do not crack so widely as those with hot dry seasons.

UDIC (1975). The moisture for the soils is such that the soil is not dry for so long as 90 days. This regime is common to many soils of humid regions.

UDOLLS (1975). More or less freely drained Mollisols of the humid mid-latitudes. They were formerly called Prairie soils (1938).

UDULTS (1975). Freely drained Ultisols low in organic matter in warm temperate or tropical regions.

ULTISOLS (1975). Soils with evidence of clay accumulation but are more leached than Alfisols. Generally they are low in mineral plant nutrients but respond to fertilizers. They are found in humid warm temperate to tropical regions. Formerly mostly included with Red-Yellow Podzolic soils (1938).

UMBREPTS (1975). Acid, dark reddish or brown, well drained Inceptisols of middle to high latitudes with good rainfall, developing under mixed grasses, shrubs, and trees.

USTALFS (1975). Alfisols with irregular summer distribution of moisture in semi-arid regions.

USTERTS (1975). Vertisols in areas of monsoon climates of tropical and subtropical areas with two rainy and two dry seasons, and of warm temperate regions with low summer rainfall. They can be well used only with heavy machinery.

USTIC. In ustic climates the soils have moisture regimes between those of aridic and udic areas.

USTOLLS (1975). Well drained Mollisols of middle and low latitudes with limited rainfall mainly in the growing season. Mostly developed under grass. Drought in some years.

USTOX (1975). Oxisols that are mostly red and are dry for long periods, with about 90-day rainy seasons, in tropical regions. Some crops can be grown without irrigation. They are developed under savanna and thin forest.

USTULTS (1975). Well drained Ultisols of warm regions of high rainfall and a pronounced dry season. The vegetation is commonly savanna that has encroached thin forest after fires. They have reddish or yellow surface soils with more clayey horizons beneath.

VERTISOLS (1975). These soils are developed from clays that swell and shrink with changes in moisture. With time they move so much that they have gilgai microrelief. They are most common in warm temperate to tropical regions.

XERALFS (1975). Alfisols in areas with cool moist winters and warm dry summers.

XERIC SOILS (1975). Soils in areas with moist cool winters and warm dry summers.

XEROLLS (1975). Mollisols with dry summers, and cool, moist winters. Their moisture regimes border those of Aridisols.

Literature cited

ACHESON, DEAN. 1969. *Present at the Creation.* W. W. Norton, New York. 798 p.

AGRAWAL, G. D. 1970. *Do Land Settlement Projects Provide an Economical Way to Develop Agriculture?* In *Land Reform* No. 1. FAO, Rome. p. 83–87.

AHN, PETER M. 1961. *Soils of the Lower Tano Basin, South-Western Ghana.* Ghana Ministry of Food and Agriculture, Kumasi. 265 p.

AHN, PETER M. 1970. *West African Soils.* Vol. 1. 3rd edition. Oxford University Press, U. K.

ALEXANDER, L. T., and J. G. CADY. 1962. *Genesis and Hardening of Laterite in Soils.* Tech. Bull. No. 1282. U.S. Dep. of Agriculture, Washington. 90 p.

ANDRESKI, STANISLAW. 1966. *Parasitism and Subversion: The Case of Latin America.* Random House, New York. 303 p.

ANDRESKI, STANISLAW. 1968. *The African Predicament.* Atherton, New York. 237 p.

BALDWIN, ROBERT E. 1966. *Economic Development and Export Growth: A Study of Northern Rhodesia 1920-60.* University of California Press, Berkeley. 254 p.

BARRACLOUGH, SOLON. 1969. *Why Land Reform?* In *Ceres.* 2(2):21–24. FAO, Rome.

BARTLETT, HARLEY H. 1955-1961. *Fire in Relation to Primitive Agriculture and Grazing in the Tropics: Annotated Bibliography.* Vol. I, 1955, 568 p.; Vol. II, 1957, 873 p.; and Vol. III, 1961, 216 p. (Processed) University of Michigan, Ann Arbor.

BAUER, P. T. 1965. *Economic Analysis and Policy in Under-developed Countries.* Routledge and Kegan Paul, London. 143 p.

BINNS, B. O. et al. 1950. *The Consolidation of Fragmented Agricultural Holdings.* FAO Agricultural Studies No. 11, Washington-Rome. 99 p.

BRADFIELD, RICHARD. 1969. *Training Agronomists for Increasing Food Production in the Humid Tropics.* p. 45–63. In *International Agronomy.* Special Pub. No. 15, American Society of Agronomy, Madison, Wisconsin.

BRADFORD, WILLIAM. 1856. *History of Plymouth Plantation, 1606–1646.* Little, Brown and Co., Boston. 476 p. (Or any other standard edition.)

BROEKMEIJER, M. W. J. M. 1966. *Fiction and Truth About "The Decade of Development."* Leyden, The Netherlands. 151 p. (paper).

BROUSEK, ING. JIŘI. 1971. *Influence of Economic Changes in Rural Planning in Czechoslovakia.* In *World Crops.* 23:252–255. London.

BUCKMAN, H. O., and NYLE C. BRADY. 1974. *The Nature and Properties of Soils.* 8th ed. Macmillan, New York. 665 p.

BYERS, HORACE G., et al. 1938. *Selenium Occurrence in Certain Soils in the United States with a Discussion of Related Topics.* Tech. Bull. No. 601. U.S. Dep. of Agriculture, Washington. 75 p.

CANADA DEP. OF AGRICULTURE. 1971. *Canada and FAO.* Publication No. 1435, Ottawa. 103 p.

CARROLL, LEWIS. [pseud.] 1872. *Through the Looking Glass.* Oxford University Press (New York, 1971). (Or any standard edition)

CARROLL, PAUL H. and R. C. MALMGREN. 1967. *Soil Survey of Southwest Region of the Republic of Ivory Coast.* (Processed) Development and Resources Corporation, New York. 2 vol.

CHARTER, C. F. 1948. *Methods of Soil Survey in Use in the Gold Coast.* In *Comptes Rendus, Conférence Africaine des Sols, Goma, Congo Belge.* Brussels, Belgium. p. 109–120.

CIC-AID. 1968. *Building Institutions to Serve Agriculture.* Purdue University, Lafayette, Ind. 236 p.

COOMBS, PHILIP H. et al. 1971. *Education on a Treadmill* and other essays. In *Ceres.* 4(3):23–51. FAO, Rome.

COONEY, S. 1968. *The East African Scientific Literature Service.* The Agricultural Institute, Dublin, Ireland. 66 p.

CRAWFORD, JOHN MARTIN. 1888. *The Kalevala,* the Epic Poem of Finland. (Transl. from Finnish.) Vol. I, p. 22, 24. John B. Alden, New York.

CROWTHER, FRANK. 1948. *Agricultural Investigation in the Sudan.* Oxford University Press, London. p. 416–940.

DALRYMPLE, D. G. 1971. *Survey of Multiple Cropping in Less Developed Nations.* Bulletin FEDR-12, U.S. Dep. of Agriculture, Washington. 108 p.

DARLINGTON, C. D. 1969. *The Evolution of Man and Society.* George Allen and Unwin, London. 753 p.

DE SCHLIPPE, PIERRE. 1956. *Shifting Cultivation in Africa: The Zande System.* Routledge and Kegan Paul, London. 504 p.

DONAHUE, R. L., J. C. SCHICKLUNA, and L. S. ROBERTSON. 1971. *Soils: An Introduction to Soils and Plant Growth.* 3rd ed. Prentice-Hall Inc. Englewood Cliffs, New Jersey. 587 p.

DOST, H. (ed.). 1973. *Acid Sulphate Soils.* International Institute for Land Reclamation and Improvement. Pub. 18. Wageningen, The Netherlands. Vol. I, 295 p. and Vol. II, 406 p.

DUCHAUFOUR, P. et al. 1970. *Précis de Pedologie*. Mason, Paris. (An edition in English is planned.) 481 p.

DUCKWORTH-BARKER, V. 1970. *Breakthrough to Tomorrow*. United Nations, New York. 72 p.

ECHEVERRIA-ALVAREZ, LUIS. 1971. *We Need Economic Growth—to Share Profits with Workers and Peasants*. In *Ceres* 4(5):25-27. FAO, Rome.

EDELMAN, C. H. 1950. *Soils of the Netherlands*. North Holland Publishing Co., Amsterdam. 177 p.

EDELMAN, C. H., and J. M. VAN STAVERN. 1958. *Marsh Soils in the United States and The Netherlands*. In *Journal of Soil and Water Conservation*. 13:5-17.

FAO. 1946. *Report of the First Session of the Conference, 1945*. FAO, Washington. 89 p.

FAO. 1963. *Possibilities of Increasing Output of Livestock Production*. In *Possibilities of Increasing World Food Production*. Rome. p. 147-187.

FAO. 1965. *Soil Erosion by Water*. Agricultural Development, Paper No. 81. Rome. 284 p.

FAO. 1970. *Provisional Indicative World Plan for Agricultural Development*. In two volumes. Rome. 672 p.

FAO. 1971. *Land Reform, Land Settlement, and Cooperatives*. (Processed) Rome. Vol. 1, 104 p.

FAO. 1969-71. *The State of Food and Agriculture*. 1969, 234 p.; 1970, 274 p.; 1971, 234 p. Rome.

FAO-UNESCO. 1971. *Soil Map of the World, Vol. IV. South America*. Illus., two large maps at 1:5,000,000. Rome. 193 p.

FERBER, A. E. 1969. *Windbreaks for Conservation*. Agricultural Information Bull. 339. U.S. Dep. of Agriculture, Washington. 30 p.

FORD FOUNDATION AGRICULTURAL PRODUCTION TEAM. 1959. *Report on India's Food Crisis and Steps to Meet It*. The Government of India. New Delhi. 258 p.

FRANK, ANDRE GUNDER. 1969. *Latin America: Underdevelopment or Revolution*. Monthly Review Press, New York. 409 p.

FREE, E. E. 1911. *The Movement of Soil by the Wind*. (with a Bibliography of Eolian Geology by S. C. Stuntz and E. E. Free.) Bureau of Soils Bull. No. 68. U.S. Dep. of Agriculture, Washington, D.C. 272 p.

GORDON, J. 1969. *Extension in Developing Countries*. In *World Crops*. 21:215-216. London.

GREENE, H. 1948. *Using Salty Land*. FAO Agricultural Studies No. 3, Rome. 49 p.

HALLET, JEAN-PIERRE. 1965. *Congo Kitabu*. Hamish Hamilton, London. 404 p.

HAMBIDGE, GOVE. 1955. *The Story of FAO*. D. Van Nostrand, New York. 303 p.

HARLER, C. R. 1971. *Tea Soils*. In *World Crops*. 23(5):275. London.

HAYES, W. A. 1971. *Mulch Tillage in Modern Farming.* Leaflet No. 554.
U.S. Dep. of Agriculture, Washington. 7 p.

HEALY, W. B. et al. 1970. *Ingested Soil as a Source of Microelements for
Grazing Animals.* New Zealand Journal of Agricultural Research, 13(3):
503-521. Wellington, N. Z.

HODGES, CARL N. 1969. *Food Factories in the Desert: Accomplishments.*
In *Proceedings, Agricultural Research Institute.* National Research
Council, Washington. 113-124 p.

IGNATIEFF, V., and H. J. PAGE. 1958. *Efficient Use of Fertilizers.* FAO
Agricultural Studies No. 43. Rome. 2nd ed. 355 p. (paper). (Also
editions in Spanish and French.)

JOHNSON, HARRY G. 1967. *Economic Policies Toward Less Developed
Countries.* Brookings Institution, Washington. 279 p.

JURION, F., and J. HENRY. 1967; 1969. *De l'Agriculture Itinérante a
l'Agriculture Intensifiée.* (1967) 498 p. and English ed. (1969) *Can
Primitive Farming be Modernized?* INEAC, Brussels. 457 p. (Very
good for the Tropics.)

KELLOGG, CHARLES E. 1941. *The Soils that Support Us.* Macmillan,
New York. 370 p.

KELLOGG, CHARLES E. 1950. *Food, Soil, and People.* Manhattan Publ.
Co., New York. (For UNESCO). 64 p.

KELLOGG, CHARLES E. 1960. *Transfer of Basic Skills of Food Production.* In *The Annals of Political and Social Science.* 331:32-38.
Philadelphia, Pa.

KELLOGG, CHARLES E. 1962. *Interactions in Agricultural Development.*
In *Science, Technology and Development*—U.S. Papers Prepared for the
UN Conference on the Application of Science and Technology for the
Benefit of the Less-Developed Areas, Geneva, 1963. Vol. III, Agricul-
ture, p. 12-24. U.S. Gov. Printing Office, Washington.

KELLOGG, CHARLES E. 1964. *Potentials for Food Production.* In *Farmer's World.* U.S. Dep. of Agriculture Yearbook. Washington. p. 57-59.

KELLOGG, CHARLES E. 1971. *Town-Country Planning.* U.S. Dep. of
Agriculture Yearbook. Washington. p. 28-33.

KELLOGG, CHARLES E., and FIDELAI DAVOL. 1949. *An Exploratory
Study of Soil Groups in the Belgian Congo.* Serie Scientific No. 46.
INÉAC, Brussels. 73 p.

KELLOGG, CHARLES E., and DAVID C. KNAPP. 1966. *The College of
Agriculture: Science in the Public Service.* McGraw-Hill, New York.
237 p.

KELLOGG, CHARLES E., and A. C. ORVEDAL. 1969. *Potentially Arable
Soils of the World and Critical Measures for their Use.* In *Advances in
Agronomy.* 21:109-170. Academic Press, New York.

KENNAN, GEORGE F. 1967. *Memoirs: 1925-1950.* Little, Brown and
Co., Boston. 583 p.

KEYNES, JOHN MAYNARD. 1965. *The General Theory of Employment,
Interest and Money.* Macmillan, London. 403 p. (Or any standard
edition.)

KUBOTA, J. et al. 1967. *Selenium in Crops in the United States in Relation to Selenium-Responsive Diseases in Animals.* In *Journal of Agriculture and Food Chemistry.* 15:448–453.

LEAMY, M. L. and W. P. PANTON. 1966. *Soil Survey Manual for Malayan Conditions.* Bull. No. 119, Division of Agriculture, Kuala Lumpur, Malaysia. 226 p.

LEVY-BRUHL, LUCIEN. 1966. *How Natives Think.* Translation by Lilian A. Clare from French ed. of 1910. Washington Square Press, New York. 355 p. (paper).

LILIENTHAL, DAVID E. 1944. *TVA—Democracy on the March.* Harpers, New York. 248 p.

LLERAS-RESTREPO, C. 1971. *In Land Reform.* No. 1, FAO, Rome. p. 75.

MARTIN, KURT, and JOHN KNAPP. 1967. *The Teaching of Development Economics.* Frank Cass, London. 238 p.

MC LUHAN, MARSHALL. 1962. *The Gutenberg Galaxy.* Routledge and Kegan Paul, London. 294 p.

MICHIGAN DEPARTMENT OF STATE HIGHWAYS. 1970. *Field Manual of Soil Engineering.* 5th ed. Lansing, Mich. 474 p.

MOORMANN, F. R. et al. 1961. *Research on Acid Sulphate Soils and Their Amelioration by Liming.* Ministry of Rural Affairs. Saigon, Viet Nam. 82 p.

MORTVEDT, J. J. (ed.) 1972. *Micronutrients in Agriculture.* Soil Science Society of America, Madison, Wis. 666 p.

MOSEMAN, A. H. (ed.) 1964. *Agricultural Sciences for the Developing Nations.* American Association for the Advancement of Science Publication No. 76. Washington. 211 p.

MOSHER, A. T. 1966. *Getting Agriculture Moving.* The Agricultural Development Council, New York. 190 p.

MYRDAL, GUNNAR. 1968. *Asian Drama: An Inquiry into the Poverty of Nations.* Random House, New York. 2,284 p. in three vols. (paper).

MYRDAL, GUNNAR. 1971. *Interview.* In *Ceres* Vol. 4(2):31–34. FAO, Rome.

MYRDAL, GUNNAR. 1972. *Political Factors in Economic Assistance.* In *Scientific American.* 226:15–21. New York.

NASH, MANNING. 1966. *Primitive and Peasant Systems.* Chandler, San Francisco. 166 p.

NELSON, NORTON et al. 1971. *Hazards of Mercury.* In *Environmental Research* 4:1–69. Academic Press, New York.

NYE, P. H., and D. J. GREENLAND. 1960. *The Soil under Shifting Cultivation.* Commonwealth Bureau of Soils. Tech. Comm. No. 51. Harpenden, U.K. 156 p.

OLSON, R. S. (ed.) 1971. *Fertilizer Technology and Use.* 2nd edition. Soil Science Society of America. Madison, Wis. 611 p.

PENNY, DAVID H. 1966. *The Economics of Peasant Agriculture: The Indonesian Case.* In *Bulletin of Indonesian Studies.* No. 5, p. 1–44. Australian National University, Canberra.

POWER, JONATHAN. 1972. *The Last Thing Africa Needs is Cities.* In *Intellectual Digest* 3:22–23. Delmar, California, and also in *Commonweal,* August, 1972.

PRESIDENT'S SCIENCE ADVISORY COMMITTEE (PSAC). 1967. *The World Food Problem.* Vol. I, 127 p.; Vol. II, 772 p. The White House, Washington.

RADWANSKI, S. A. 1971. *East African Catenas in Relation to Land Use and Farm Planning.* In *World Crops,* 23:265–273. London.

RASHDALL, HASTINGS. 1936. *The Universities of Europe in the Middle Ages.* Revised ed. Oxford University Press. 3 vols.

RICHARDS, L. A. et al. 1954. *Saline and Alkali Soils.* Agricultural Handbook No. 60. U.S. Dep. of Agriculture, Washington. 160 p.

ROBINSON, JOAN. 1971. *Economic Heresies: Some Old-Fashioned Questions in Economic Theory.* Macmillan, London. 150 p.

ROBINSON, RONALD. 1971. *Developing the Third World: The Experiences of the Nineteen-Sixties.* Cambridge University Press, London and New York. 289 p.

ROBNETT, O. L., and M. R. THOMPSON. 1970. *Rcommendations for Stabilization of Illinois Soils.* (Processed) Univ. of Illinois Engineering Exp. Sta. Bull. No. 502. Urbana. 225 p.

ROLL, ERIC. 1968. *The World after Keynes.* Praeger, New York. 193 p.

RUFFIN, EDMUND. 1832. *Calcareous Manures.* J. W. Campbell, Petersburg, Virginia. 242 p. (Four subsequent editions to 1852.)

RUHE, ROBERT V. 1956. *Landscape Evolution in the High Ituri, Belgian Congo.* Série Scientific No. 66. INÉAC, Brussels. 108 p.

RUHE, ROBERT V. et al. 1967. *Landscape Evolution and Soil Formation in Southwestern Iowa.* Tech. Bull. 1349. U.S. Dep. of Agriculture, Washington. 242 p.

RUSSELL, E. JOHN, and E. WALTER RUSSELL. 1961. *Soil Conditions and Plant Growth.* 9th ed. Longmans, London. 688 p. (New edition in press.)

SCHULTZ, C. B., and J. S. FRYE (ed.). 1968). *Loess and Related Eolian Deposits of the World.* Univ. of Nebraska Press, Lincoln. 369 p.

SCHULTZ, T. W. 1964. *Transforming Traditional Agriculture.* Yale University Press, New Haven. 212 p.

SKOROPANOV, S. G. 1961. *Reclamation and Cultivation of Peat-Bog Soils.* Translation from edition in Russian, Minsk, 1961 by N. Kaner. Israel Program for Scientific Translations, Jerusalem, 1968. 239 p.

SMITH, ADAM. 1762. *The Wealth of Nations.* The Modern Library (New York, 1965.) (Or any standard edition)

SMITH, GUY D. 1965. *Lectures in Soil Classification.* In *Pedologie.* Special issue No. 4, Ghent, Belgium. 134 p.

SNOW, C. P. 1959. *The Two Cultures and the Scientific Revolution.* (The Rede Lectures) Cambridge University Press, London. 51 p.

SOIL SURVEY STAFF. 1951. *Soil Survey Manual.* U.S. Dep. of Agriculture Handbook No. 18. Washington. 503 p.

SOIL SURVEY STAFF. 1960. *Soil Classification: A Comprehensive System, 7th Approximation.* U.S. Dep. of Agriculture, Washington. 265 p. (A new edition in preparation.)

SOIL SURVEY STAFF. 1975. *Soil Taxonomy: A Basic System of Soil Classification for Making and Interpreting Soil Surveys.* Agriculture Handbook no. 436. U.S. Government Printing Office, Washington, D.C. (in press).

STAKMAN, E. C., RICHARD BRADFIELD, and PAUL C. MANGELSDORF. 1967. *Campaigns Against Hunger.* Harvard Press, Cambridge. 328 p.

STEWART, A. B. 1947. *Report on Soil Fertility Investigation in India with Special Reference to Manuring.* Army Press, Delhi, India. 158 p.

STEWART, MICHAEL. 1967. *Keynes and After.* Penguin Press. Harmondsworth, U.K. 271 p. (paper).

TEPICHT, J. 1969. *Problems of the Restructuring of Agriculture in the Light of the Polish Experience.* p. 534–554. In Ugo Papi and Charles Nunn (ed.) *Economic Problems of Agriculture in Industrial Societies.* St. Martin's Press, New York.

THORP, WILLARD L. 1971. *The Reality of Foreign Aid.* Praeger, New York. 370 p.

TOTHILL, J. D. et al. 1948. *Agriculture in the Sudan.* Oxford University Press, London. 974 p.

TURNBULL, COLIN M. 1961. *The Forest People.* Chatto and Windus, London. 450 p.

UNDERWOOD, E. J. 1971. *Trace Elements in Human and Animal Nutrition.* 3rd ed. Academic Press, New York and London. 543 p.

UNITED NATIONS. 1943. *Conference on Food and Agriculture: Final Act and Section Reports.* U.S. Government Printing Office, Washington, D.C. 61 p.

UNITED NATIONS. 1963. *Science and Technology for Development.* Vol. III, *Agriculture.* New York. 309 p.

UNITED NATIONS. 1970. *International Development Strategy.* New York 20 p.

U.S. DEP. OF AGRICULTURE. 1938. *Soils and Men,* Yearbook. Washington, D.C. 1232 p.

U.S. DEP. OF AGRICULTURE. 1957. *Soil,* Yearbook (Management). Washington, D.C. 784 p.

U.S. DEP. of HEALTH, EDUCATION, AND WELFARE. 1969. *Report of the Secretary's Commission on Pesticides and their Relationship to Environmental Health.* Washington, D.C. 677 p.

VIETS, FRANK G., and R. H. HAGEMAN. 1971. *Factors Affecting the Accumulation of Nitrate in Soil, Water, and Plants.* U.S. Dep. of Agriculture Handbook No. 413. Washington, D.C. 63 p.

WALLACE, T. 1951. *The Diagnosis of Mineral Deficiencies in Plants by Visual Symptoms: A Colour Atlas and Guide.* 2nd ed. H.M.S., London. 107 p.

WALSH, T., P. RYAN, and STAFF. 1969. *West Donegal Resource Survey.* The Agricultural Institute, Dublin, Ireland. Part 1, 78 p.; Part 2, 86 p.; Part 3, 128 p.; and Part 4, 59 p.

WASHINGTON, GEORGE. 1803. *Letters from his Excellency George Washington to Arthur Young and Sir John Sinclair, containing an Account of his Husbandry, etc.* Cotton and Stewart, Alexandria and Fredericksburg, Virginia. 128 p.

WEAVER, J. E. 1968. *Prairie Plants and their Environment. A Fifty-Year Study in the Midwest.* University of Nebraska Press, Lincoln. 276 p.

WELLHAUSEN, E. J. et al. 1970. *Strategies for Increasing Agricultural Production on Small Holdings. (Puebla Project).* CIMMYT, Mexico, D.F. 86 p.

WELLS, ORIS V. 1971. *International Agricultural Adjustments.* In *American Journal of Agricultural Economics.* 53:786–792.

WILLIAMS, M. S., and J. W. COUSTON. 1962. *Crop Production Levels and Fertilizer Use.* FAO, Rome. 48 p.

WHYTE, R. O. 1962. *The Myth of Tropical Grasslands.* In *Tropical Agriculture.* Vol. 39, p. 1–11. Trinidad, West Indies.

WOOD, ALAN. 1950. *The Groundnut Affair.* Bodley Head, London. 264 p.

WRIGHT, A.C.S. 1963. *Soils and Land Use of Western Samoa.* Soil Bureau Bull. No. 22, Wellington, N.Z. 191 p.

ZUCKERMAN, LORD. 1971. *Scientific Expectations and Disappointments.* The Times (London) Literary Supplement. 29 October 1971, p. 1439–1452.

Index